Filipe Ijuim

DSC Configuration Using the PSIM/SimCoder Tool

Filipe Ijuim

DSC Configuration Using the PSIM/SimCoder Tool

Application to Switched Converters

ScienciaScripts

Imprint

Cover image: www.ingimage.com

This book is a translation from the original published under ISBN 978-620-6-75945-4.

Publisher:
Sciencia Scripts
is a trademark of
Dodo Books Indian Ocean Ltd. and OmniScriptum S.R.L publishing group

120 High Road, East Finchley, London, N2 9ED, United Kingdom
Str. Armeneasca 28/1, office 1, Chisinau MD-2012, Republic of Moldova, Europe
Managing Directors: Ieva Konstantinova, Victoria Ursu
info@omniscriptum.com

Printed at: see last page
ISBN: 978-620-8-38453-1

Contents

SUMMARY

IJUIM, Filipe K. Configuration of a Digital Signal Controller Applied to Switched Converters Using the PSIM/SimCoder Tool. 2014. TCC (Bachelor's Degree in Electrical Engineering - Area: Electronics and Microprocessors) - Santa Catarina State University, Joinville, 2014.

This final project presents the study of elements of the simulation software PSIM® , used to generate codes in C language. The SimCoder blocks can be configured and used in switching converter simulations. The generated code can be imported into the Code Composer Studio v5.1 programme, without having to make any modifications or adjustments to the code. SimCoder presents these blocks for implementing digital control for a Digital Signal Controller (DSC) from Texas Instruments, the TMS320F28335, a floating point controller. Using the SimCoder blocks, you can configure PWM outputs, analogue inputs, digital inputs and outputs, among other options found in the DSC. After studying switched converters, simulations are carried out with the blocks mentioned, codes are generated and tests are carried out with the DSC and the NPCm, a three-level bidirectional converter. Finally, the results obtained experimentally are presented, showing the waveforms of the DSC and the converter used, verifying the operation of the codes generated.

Keywords: PSIM. Digital Signal Controller. Switched converters. Digital Control.

CHAPTER 1

1. INTRODUCTION

In converter design, an extremely important part is the command design. For very complex converters, equally complex controls are required, taking up as much design time as other stages such as research, converter design and implementation. They are normally used to implement microcontrollers, Digital Signal Processors (DSP), Digital Signal Controllers (DSC), FPGA's, among others. Among the factors that determine the choice of controller, we can highlight its programming language, processing time, maximum clock frequency, type of memory, number of inputs and outputs available, among other factors.

Another important part of converter design is simulation. Circuit simulation software is often used for this process. By testing the simulations, you can get an estimate of how the converter will behave, often very close to the actual behaviour obtained.

Both stages are extremely important for a project, requiring time for research, time to learn the software, time to study the programming language, to master the microcontrollers, and then to carry out the project itself.

Eventually, the components of this project, i.e. the simulation software and the microcontroller, have tools that help in the process, making it possible to drastically reduce the time it takes to realise the project. These tools also enable rapid learning, making it easier to use the simulation software and the controller.

This paper will explore one of these tools, the PSIM simulation software® , specifically the SimCoder elements. After simulating circuits, this tool allows code to be generated in C language, which can be implemented on a Texas Instruments DSC, the TMS320F28335. To this end, all the elements present in SimCoder will be studied, as well as other elements of PSIM® that can be used together. As there are no scientific articles on this tool, user manuals, examples and the software's own help tool will be used as bibliographies.

Once the code has been generated, it needs to be understood, so a study is made of the C language, the DSC used and Code Composer Studio, the software responsible for compiling the code and passing it on to the DSC. The bibliography includes user manuals, application notes and examples.

In order to understand and test this tool, studies and simulations will be carried out on some switched converters, enabling the SimCoder elements presented to be used. The converters chosen for the study are power circuits that have silicon switches such as MOSFET and IGBT transistors. These circuits convert electrical energy from direct current to direct current (DCCC) or from direct current to alternating current (DC-AC). Books on power electronics (BARBI, 2006), (MARTINS and BARBI, 2006) will be used for this study.

Finally, the generated codes will be tested on the converters studied. Didactic models found in the laboratory will be used. The aim is to verify the effectiveness of the simulation process, code generation and control implementation in the DSC, presenting the results obtained.

CHAPTER 2

2. BASIC ELECTRONIC CONVERTERS

Power electronics can be defined as an applied science dedicated to the study of static electrical energy converters (BARBI, 2006). Static converters are systems capable of performing some kind of electrical energy treatment, featuring passive elements, i.e. resistors, inductors and capacitors, and active elements that perform some kind of interruption, such as diodes, thyristors and transistors.

The main functions of static converters include: Rectifier, capable of transforming alternating current into direct current; Direct Frequency Converter, capable of converting alternating voltage into voltage of a different frequency; Inverter, in charge of transforming direct current into alternating current and DC-DC Converter, capable of changing the level of direct current. This work will exclusively study DC-DC converters and inverters, also known as DC-DC converters. Among the DC-DC converters, the Buck converter will be studied, while among the DC-AC converters, the full bridge converter and a three-phase converter will be studied.

2.1. BUCK CONVERTER

The Buck converter is used to convert an average DC voltage value at the input into a lower average DC voltage value at the output, i.e. it is a voltage step-down converter. This converter is used to control the flow of energy between two direct current systems.

As shown in Figure 1, the Buck converter has some components that define its operation. Switch S has the function of switching, determining two stages of operation.

In the first stage, when the switch is closed, the circuit is powered by the voltage source v_g. In the second stage, with the switch open, another element will be responsible for supplying the load, the inductor L, which works in the circuit as a current source, releasing the energy stored during the first stage.

Figure 1 - Basic Buck Converter topology

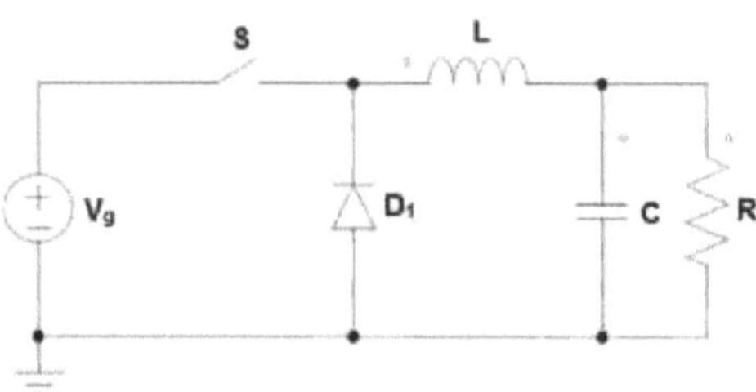

Source: Author's own production.

If the inductor current reaches zero before the end of the switching period, the converter is considered to be operating in discontinuous conduction, otherwise conduction is continuous. Figure 2 shows the path taken by the current during the first stage.

Figure 2 - First operating stage of the Buck converter.

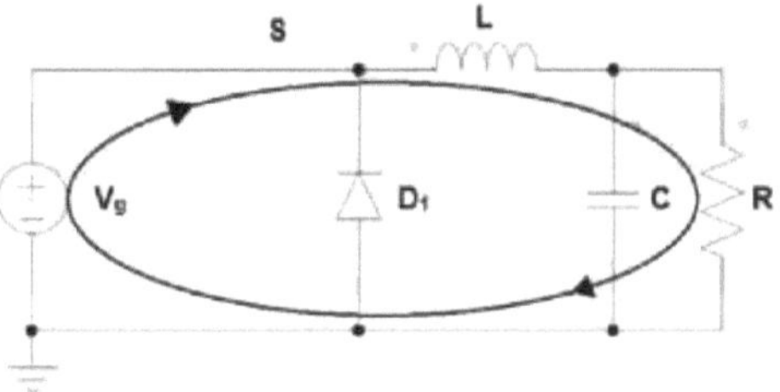

Source: Author's own production.

Diode D1 performs two functions, depending on the stage the circuit is in. In the first stage, with the switch closed, the diode is responsible for blocking any possible current contrary to the current imposed by the source. With the switch open, i.e. in the second stage, the diode freewheels, allowing the current from inductor L to flow through the load. Figure 3 shows how the Buck converter works during the second stage.

Figure 3 - Second stage of Buck converter operation.

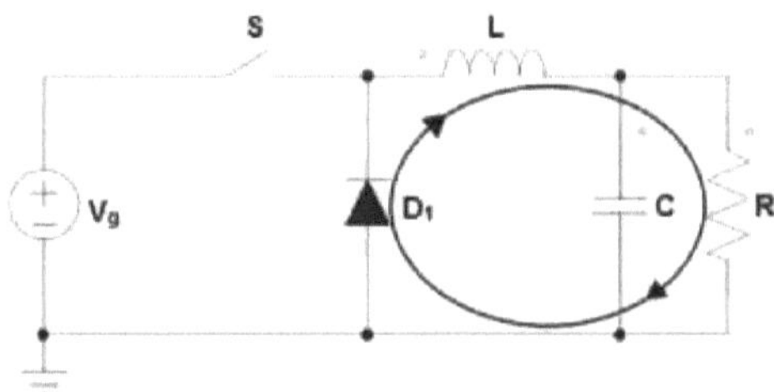

Source: Author's own production.

Operating in continuous conduction mode, the Buck converter has a lower effective current than in discontinuous conduction mode, making continuous conduction a better option. The main waveforms of Buck converter operation in continuous conduction can be seen in Figure 4.

Figure 4 - Buck converter waveforms: a) Diode voltage; b) Inductor current; c) Switch current; d) Diode current.

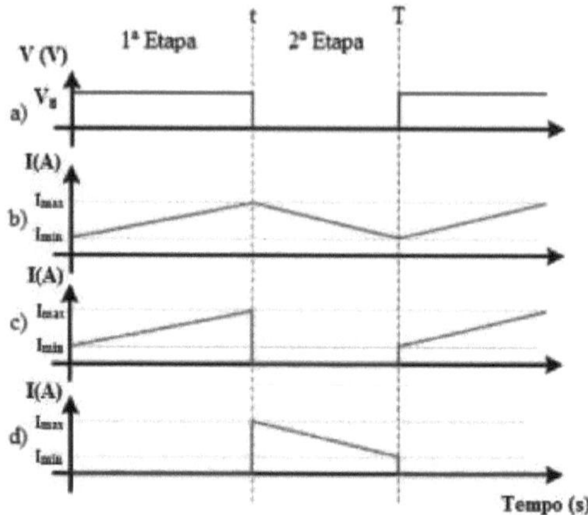

Source: Author's own production.

Note that the waveforms shown are found when the converter is in steady state. The waveforms shown show that T is the switching period and t is the time the switch remains closed. The ratio between the two times mentioned is called the cyclic ratio, or duty cycle. You can also see that the currents have a maximum and minimum value. For discontinuous conduction, the minimum current value would be zero.

From simulations in the PSIM software, a characteristic similar to a second-order system can be observed, as shown in Figure 5.

Figure 5 - Buck converter output voltage.

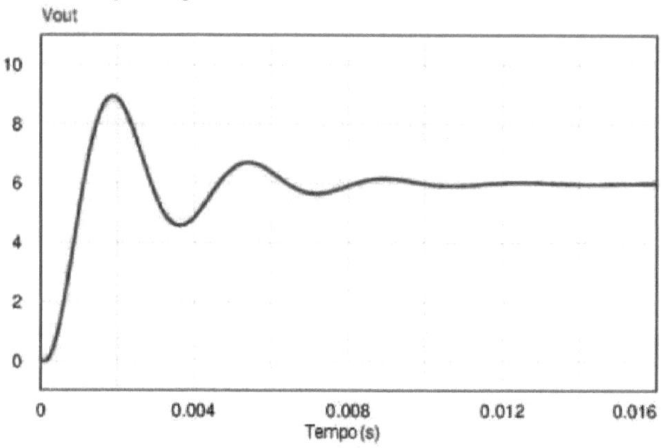

Source: Author's own production.

Due to this observed characteristic of the Buck converter, it is common to find control designs with the aim of correcting or cancelling the observed voltage peak or reducing the response time. To design the controller, it is necessary to obtain the mathematical model of the converter. To do this, consider the Buck converter simplification shown in Figure 6.

Figure 6 - Simplified Buck converter model.

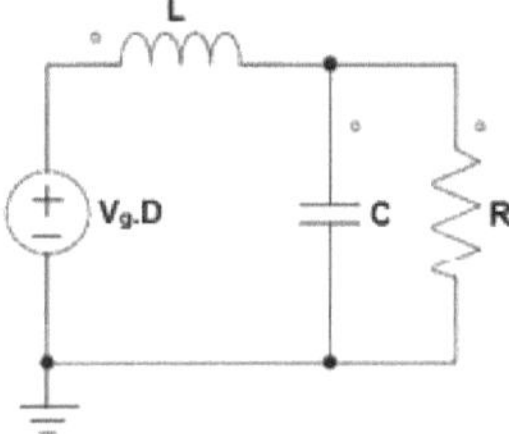

Source: Author's own production.

The voltage source of the circuit corresponds to the average voltage of the diode, where *Vg* is the input voltage and *D* is the cyclic ratio. The following equations can be obtained from the circuit shown:

$$-V_g.D + L\frac{di_L(t)}{dt} + v(t) = 0 \quad (1)$$

$$i_L(t) = i_C(t) + \frac{v(t)}{R} \quad (2)$$

$$i_C(t) = C\frac{dv(t)}{dt} \quad (3)$$

Where i_L and i_c are the currents in the inductor and capacitor respectively, and *v(t)* is the circuit's output voltage, i.e. the voltage at the load. Applying equation (2) to equation (1), we obtain:

$$-V_g.D + L\frac{di_C(t)}{dt} + \frac{L}{R}\frac{dv(t)}{dt} + v(t) = 0 \quad (4)$$

Substituting equation (3) into equation (4) gives the following equation.

$$LC\frac{d^2v(t)}{dt^2} + \frac{L}{R}\frac{dv(t)}{dt} + v(t) = V_g.D \quad (5)$$

Considering the concept of instantaneous mean value and that the value of the cyclic ratio also varies with time, you can modify equation (6) and then apply the Laplace Transform (7).

$$LC\frac{d^2\hat{v}(t)}{dt^2} + \frac{L}{R}\frac{d\hat{v}(t)}{dt} + \hat{v}(t) = V_g.D(t) \quad (6)$$

$$LC\,s^2V(s) + \frac{L}{R}\,s\,V(s) + V(s) = V_g.D(s) \quad (7)$$

Considering F(s) as the output signal and *d(s)* as the converter's input signal, we find the Buck converter's transfer function.

$$\frac{V(s)}{D(s)} = \frac{V_g}{LC\,s^2 + \frac{L}{R}s + 1} \quad (8)$$

Using the *s-domain Transfer Function* component of the PSIM *software*, the model obtained can be simulated and compared with the converter waveform. The following values were used for this simulation:

Table 1 - Values used to simulate the Buck Converter.

vg	12V
L	600μH
C	500pF
R	2,5Ω

Source: Author's own production.

The result obtained is shown in Figure 7. It can be seen that the model behaves very closely to the behaviour of the converter.

Figure 7 - Comparison between the voltages of the switched converter and the response of the continuous model obtained.

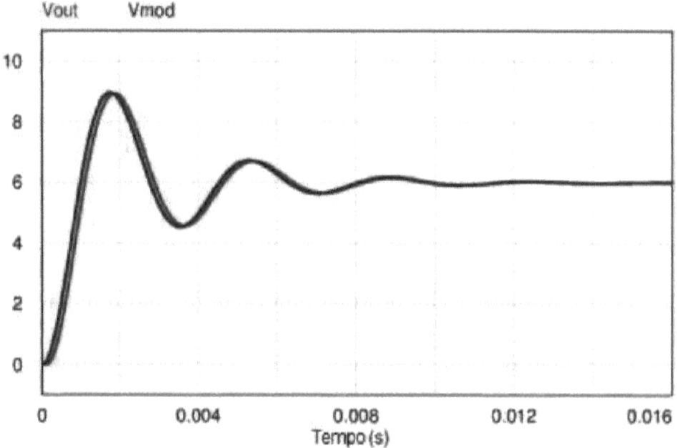

Source: Author's own production.

In the same way that the Buck converter's voltage model is obtained, the current model can be found, as shown below.

$$\frac{I_l(s)}{D(s)} = \frac{Vg}{R}\left[\frac{RCs+1}{LCs^2+\frac{L}{R}s+1}\right] \qquad (9)$$

Where *li is the* current in the inductor.

Other converters that will not be presented have commands similar to the Buck converter. Other DC-DC converters, such as the Boost converter or the Buck-Boost converter, have only one switch, similar to the Buck, so only one PWM signal is needed for control. The models for these converters can be obtained in the same way as for the Buck converter, so you can carry out simulations with the converter and design the controller.

2.2. DC-AC CONVERTERS

Inverters are static circuits (i.e. they have no moving parts) that convert DC power into AC power with the desired frequency and output voltage or current (AHMED, 2000). Therefore, this converter takes DC voltage as its input and alternating voltage as its output. To make the output voltage a sine wave, the pulse width modulation (PWM) method is used, with a sinusoidal signal as the reference signal.

There are various types of inverters, differing in the number of switches used, types of control, number of phases, etc. The most basic type is the half-bridge inverter, shown in Figure 8.

Figure 8 - Half-bridge inverter.

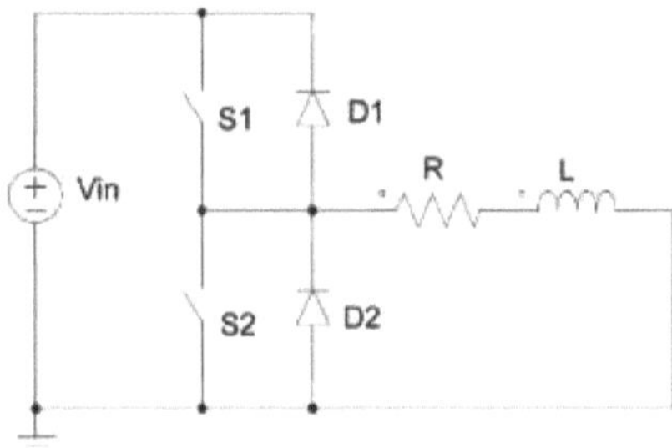

Source: Author's own production.

This converter has two switches, which have complementary commands, i.e. when one switch is activated, the other must be deactivated so that the voltage source is not short-circuited. To ensure that the switches are not activated simultaneously, the command is made with a small time difference between the end of the cycle of one switch and the start of the cycle of the other. This time is called dead time.

In the circuit, when switch S1 is conducting, the voltage at load RL is the input source voltage. When switch S2 goes into conduction, the output voltage is zero, thus obtaining a square wave. Note that in both circuits, diodes D1 and D2 have the function of conducting opposing currents. These diodes are called feedback diodes.

Another type of inverter is the full bridge inverter. This inverter, as shown in Figure 9, has four switches distributed over two arms. With this configuration, different waveforms can be obtained at the output, depending on the type of command.

Figure 9 - Full bridge inverter

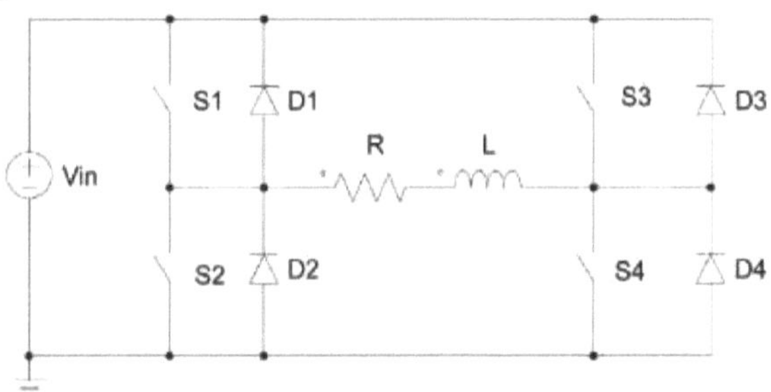

Source: Author's own production.

The table below shows the possible combinations, considering that the switches on the same arm, i.e. S1 and S2 or S3 and S4, cannot be operated simultaneously.

Table 2 - Operation of the Full Bridge Inverter.

State	SI	S2	S3	S4	Output voltage
1	On	Off	Off	On	Vin

2	On	Off	On	Off	0
3	Off	On	On	Off	-Vin
4	Off	On	Off	On	0

Source: Author's own production, based on Ahmed, Power Electronics (AHMED, 2000).

One control option for the full bridge inverter is to use states 2 and 3 shown in the table. Using only one reference signal, switches S1 and S4 are controlled simultaneously with a PWM signal, and with the complement of this signal switches S2 and S3 are controlled simultaneously. This produces a square wave with a maximum voltage of Vin and a minimum voltage of -Vin.

However, this inverter can be controlled using the four states shown. Starting with state 2, with switches S1 and S4 conducting and S2 and S3 not conducting. Next, switch S3 is turned on and S4 is turned off, keeping switches S1 and S2 as they were in the previous stage, as shown in state 1. Switches S3 and S4 are then kept on, reversing the states of switches S1 and S2, referring to state 3. Finally, in state 4, switches S1 and S2 are kept on, S3 is turned off and switch S4 is turned on.

These steps produce a square wave at the load, which has three different levels. Modulation for this type of control is done with two reference sinusoids, offset by 180°. The first reference signal generates the PWM signal for switches S1 and S2, which are complementary to each other, while the second reference signal generates the PWM signal for switches S3 and S4. Figure 10 shows the inverter's output signal and its reference signals.

Figure 10 - Output voltage of the three-level Full Bridge Inverter and sinusoidal reference signals.

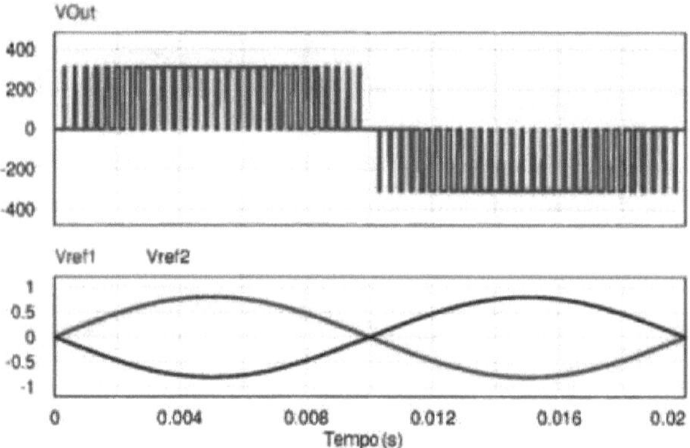

From the output of the inverter, a sine wave can be obtained by filtering out undesirable harmonics. The study of different types of inverters and their control is aimed at reducing these harmonics, thus making it possible to use simpler filters and reducing losses in the process. Less harmonics can be obtained by increasing the number of switches on the same arm of the inverter, which also makes its control more complex.

A three-phase inverter has three arms with two switches and two feedback diodes, as shown in Figure 11. Using a three-phase load bank, the output is three waveforms similar to those found in the full-bridge inverter, but offset by 120°.

Figure 11 - Three-phase inverter.

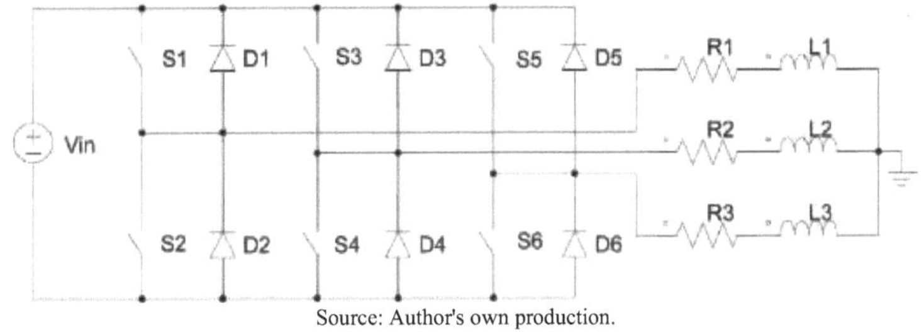

Source: Author's own production.

Six PWM signals are used to control this inverter, one signal for each switch. As with the full-bridge inverter, in the three-phase inverter the switches on the same arm must not be actuated simultaneously. The PWM signals are generated by three sinusoidal references offset by 120° from each other, compared to the same triangular carrier, as shown in Figure 12.

Figure 12 - Reference and carrier signals for controlling a three-phase inverter.

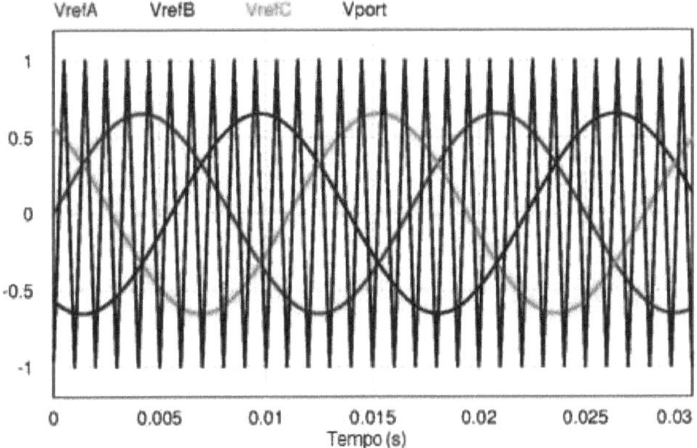

Source: Author's own production.

Another possible topology is shown in Figure 13. The circuit shown is called an NPC (Neutral Point Clamped) Inverter. This converter has three arms, with four switches on each arm, and outputs to a three-phase load.

The reason for having more switches on each arm is that this configuration allows more levels to be obtained at the output of the converter, thus simplifying the design of the filter used, due to the reduction in harmonic content, and also reducing the stress on the switches.

The control of this inverter is more complex, as twelve PWM channels are required for the twelve switches observed. Two triangular carriers and three sinusoidal reference signals are used to control the NPC inverter. Firstly, the switches are divided into odd-numbered and even-numbered switches, as shown in the figure above. For the odd-numbered switches, the sinusoidal references are compared with the positive triangular carrier, i.e. the wave that has a minimum value of zero and a maximum value of 1.

Figure 13 - Three-phase NPC inverter.

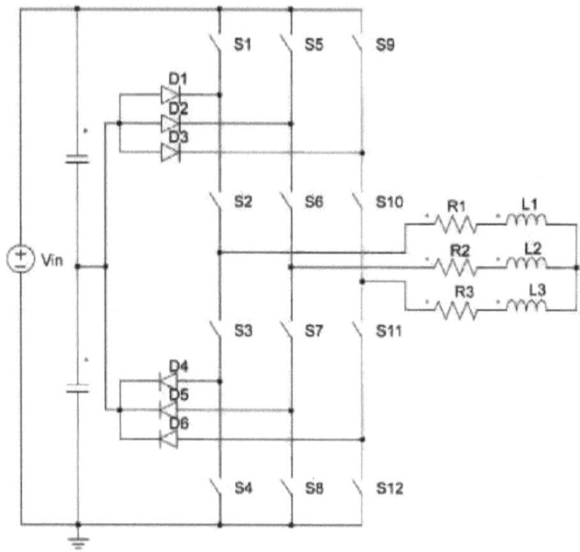

Source: Author's own production.

As with the simple three-phase inverter, a sinusoidal reference is used for each arm, offset by 120°. For switches S1 and S3, the PWM channels must be complementary, as well as for switches S5 and S7 and switches S9 and S11. The same is done for the even switches, but with the negative triangular carrier, i.e. the wave that has a maximum value of zero and a minimum value of -1. Figure 14 shows the waveforms of the reference signals and the IDP (In-Phase Disposition) modulation carriers used to control the NPC converter.

Figure 14 - NPC reference signals and carriers.

VrefA VrefB VrefC Vport1 Vport2

Source: Author's own production.

CHAPTER 3

3. SIMCODER COMPONENTS

The SimCoder tool in the PSIM software® makes it possible to create a C language code ready for implementation in a digital signal controller (DSC) from a simulation circuit. The controller in question is the TMS320F28335 from Texas Instruments, which will be studied later. SimCoder has several components, which can be found in the menu under Elements/SimCoder for CodeGeneration/TI F28335 Target. It is important to configure the simulation control before carrying out a simulation using one of the SimCoder components. Figure 15 shows the simulation control symbol and settings window.

Figure 15 - PSIM Simulation Control: (a) Symbol and (b) Parameter window

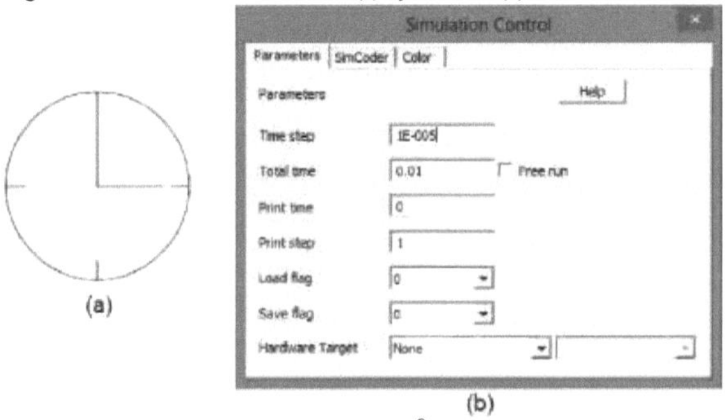

(a)

(b)

Source: PSIM .®

In Figure 15b, under Hardware Target, select the TI F28335 option, and to the right RAM Debug, to be used later in the DSC software, Code Composer Studio. The process for generating the code will be explained after the presentation of the SimCoder components.

The DSC F28335 has eighty-eight pins that can be configured as PWM outputs, digital inputs and outputs, encoders, trip-zones, counters or captures. It also has sixteen exclusive pins for A/D converters.

3.1. PWM MODULATED OUTPUTS

PWM modulation consists of transforming a reference voltage level into a square wave with a cyclic ratio proportional to this voltage. The cyclic ratio of a PWM signal is the ratio between the time the wave remains at a high level and the period of the square wave, given by:

$$D = \frac{t}{T} \qquad (10)$$

Where D is the cyclic ratio, or duty cycle, t is the time at high logic level and T is the period, which is determined as the inverse of the square wave frequency.

To generate a PWM signal, a reference signal is compared with a carrier in the shape of a triangular or sawtooth wave. The choice of carrier, amplitude and frequency depends on the application. An analogous comparison can be made with an operational amplifier, as seen in Figure 16.

Figure 16 - PWM modulator circuit.

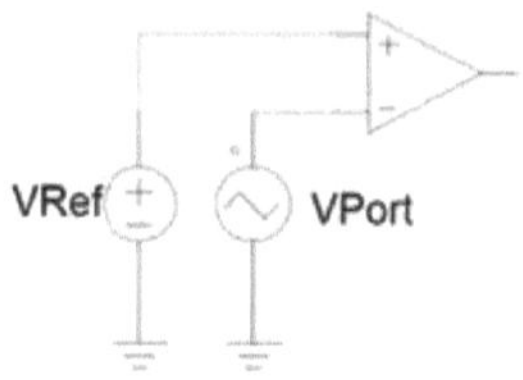

Source: Author's own production.

Depending on the application, the reference signal can be a constant or sinusoidal signal. Figure 17 shows the triangular carrier signal (VPort), the constant reference signal (VRef) and the waveform of the generated PWM signal (VPWM).

Figure 17 - Waveforms of a PWM modulation with constant reference.

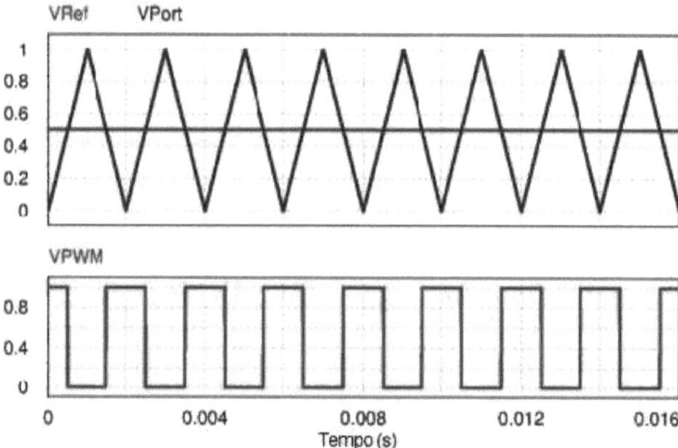

Source: Author's own production.

If the reference signal is a sine wave, the cyclic ratio varies constantly, in the same proportion as the input signal. It is important that the frequency of the carrier wave is higher than the frequency of the reference signal. Figure 18 shows the waveforms of the carrier (VPort), the sine signal (VRef), and the resulting PWM signal (VPWM).

Figure 18 - Waveforms of a PWM modulation with sinusoidal reference.

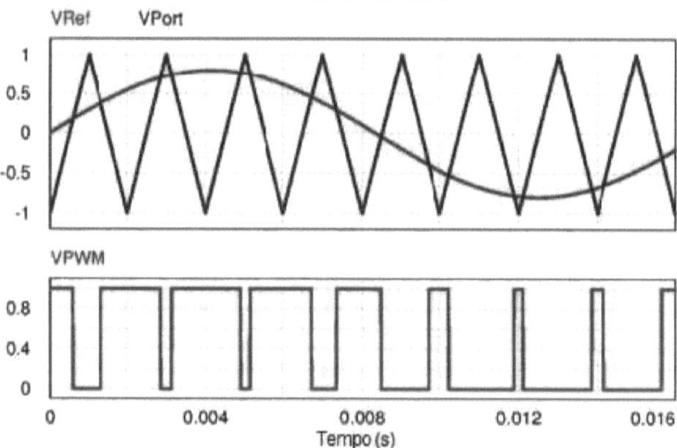

Source: Author's own production.

SimCoder has some PWM output configuration blocks. These blocks, presented below, compare an internal carrier with a reference signal. The PWM blocks found in SimCoder are: Single PWM, single-phase PWM, two-phase PWM and three-phase PWM.

3.1.1. Single PWM

The Single PWM configuration block has only one input and one output. The input must be a constant or variable signal that is between the configured peak-to-peak value. Its output will be a square wave signal with a low level equal to 0 and a high level equal to 1. Figure 19 shows the Single PWM block.

Figure 19- Single PWM block.

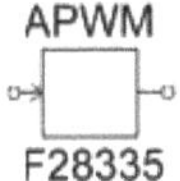

Source: Author's own production.

The configurations required for the Single PWM block, shown in Figure 6, are:

- PWM *Source:* Selection of the corresponding port to be used by the DSC. They are divided from APWM1 to APWM6, each of which has the option of selecting between two or three different GPIO ports. If you need to use more than one Single PWM generator, you must configure them with different APWMs, so you can use up to six of these blocks in one application.
- PWM *Frequency*: Definition of the frequency of the output signal, in hertz. When you open the settings window, this space will already be filled with 10 kHz.
- *Peak-to-Peak Value*: Definition of the peak-to-peak value of the input signal.
- *Offset Value*: Definition of the offset value of the input signal (DC level).

Figure 20- Configurations of the Single PWM block.

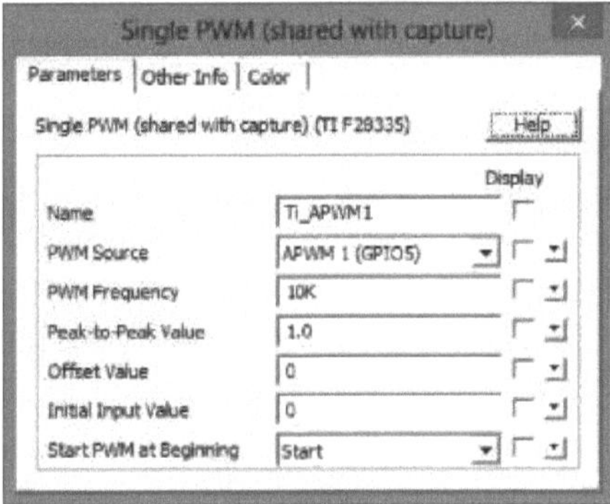

Source: PSIM.®

- *Initial Input Value:* Definition of the initial input value. This value only affects the first few cycles of the output signal.
- *Start PWM at Beginning*: Selection between "*Start*", where the PWM generator will start activated, or "*Do not Start*", where the PWM generator will depend on a *Start* PWM block to work.

The pins that can be used for this PWM generator are: GPIO1, GPIO3, GPIO5, GPIO7,

GPIO9, GPIO11, GPIO24, GPIO25, GPIO26, GPIO27, GPIO34, GPIO37, GPIO48 and GPIO49. However, you need to check in the configurations which of these ports belong to the same APWM group, as two Single PWM blocks cannot be configured for the same group.

3.1.2. Single-phase PWM

The Single Phase PWM generator also has a single input, just like the Single PWM generator, but it has two outputs, as can be seen in Figure 21. Output B consists of a complementary signal to output A.

Figure 21 - Single-phase PWM block.

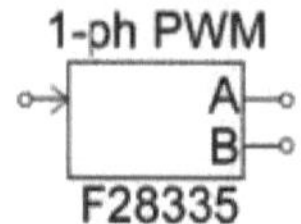

Source: Author's own production.

This block has several configuration options that the Single PWM generator does not have, allowing for a greater variety of applications. The settings window for the Single-Phase PWM module can be seen in Figure 22. Its configuration parameters are as follows:

- PWM *Source:* Selection of the corresponding ports to be used by the DSC. They are divided into PWM1 to PWM6, each of which has two GPIO ports. For PWM1, the ports are GPIO0 and GPIO1, for PWM2, the ports are GPIO2 and GPIO3, following this pattern up to PWM6. It can be seen that six single-phase generators can be used in one application.

- *Output Mode*: Definition of the output mode, between using outputs A and B, or using only output A or output B.

- *Dead Time*: Definition of the dead time between output signals, in seconds. This time is necessary in some cases so that two switches are not controlled simultaneously. When you open the settings window, this space is already filled with 4μs.

- PWM *Frequency*: Definition of the frequency of the PWM block, in hertz. When you open the settings window, this space will already be filled with 10kHz. This chosen frequency will be the sampling frequency of the A/D converters.

- PWM *Frequency Scaling Factor*: Selection between factors 1, 2 or 3. This scaling factor multiplies the PWM frequency, determining the frequency of the PWM output signal. It can be used to utilise a PWM signal with a different frequency to the sampling frequency.

- *Carrier Wave Type*: Definition of the type of carrier waveform. You can choose between triangular or sawtooth shapes.

- *ADC Trigger:* You must choose whether the PWM block will trigger the A/D Converter and which group of the converter will be triggered, group A, group B, or both.

- *ADC Trigger Position*: You must choose whether to trigger on an A/D converter, and whether to trigger on group A, B or both groups of the converter.

- *Use Trip-zones*: Definition of how *Trip-zones* 1 to 6 will be used, between disabling them, or enabling them in *One-shot* or *Cycle-by-Cycle* mode. These will be explained in the *Trip-zone* block settings.

Figure 22 - Single-phase PWM generator configurations.

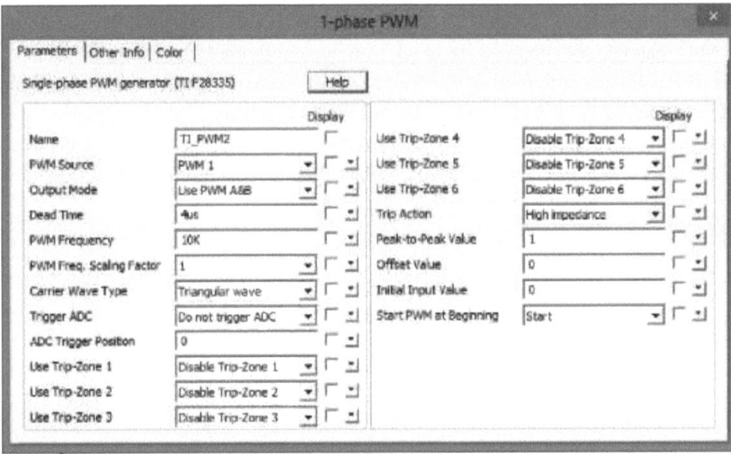

Source: PSIM.®

- *Trip Action*: Definition of how the PWM block will respond to the action of a *Tripzone*, being able to choose between high impedance for both outputs, for only output A or only B, or no action at all.
- *Peak-to-Peak Value*: Fill in the peak-to-peak value of the input signal.
- *Offset Value*: Fill in the offset value of the input signal (DC level).
- *Initial Input Value:* Fill in the initial *input* value. This value only affects the first few cycles of the output signal.
- *Start PWM at Beginning*: You must select between "*Start*", in which case the PWM generator will start activated, or "*Do not Start*", in which case the PWM generator will depend on a *Start* PWM block to work.

3.1.3 Single-phase PWM with phase shift

This PWM block is similar to the previous one, with an additional input. The *In* input is the reference signal input, also found in the Single Phase PWM block. The *Phase* input is used to phase-shift the PWM signal generated by this block in relation to an ordinary single-phase PWM block. This block must be used in conjunction with a single-phase PWM block, as it will use the PWM signal from this block as a reference. Figure 23 shows this block.

Figure 23 - Single-phase PWM block with phase change.

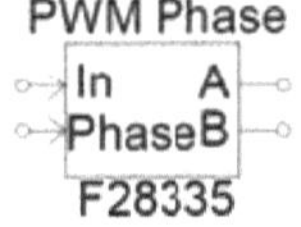

Source: Author's own production.

To use this block, the PWM channels are divided into two groups. The first group includes the PWM1, PWM2 and PWM3 channels, while the second group includes the PWM1, PWM4, PWM5 and PWM6 channels. The single-phase reference PWM block and this block must be configured for channels belonging to the same group, and the single-phase PWM block with phase shift cannot be configured for the PWM1 channel. For example, if the Single Phase PWM block has been configured for the PWM2 channel, the block with phase shift must be configured for the PWM3 channel. If the Single-Phase PWM block has been configured for the PWM4 channel, the phase-shift block can be configured for the PWM5 or PWM6 channels. If PWM1 is configured as the reference, the phase-shift block can be configured for any other channel, as the PWM1 channel belongs to both groups.

The configuration parameters for this component, with the exception of the channel to be chosen, are the same as those shown for the Single-Phase PWM block, as seen in Figure 24.

Figure 24 - Single-phase PWM block configurations with phase change.

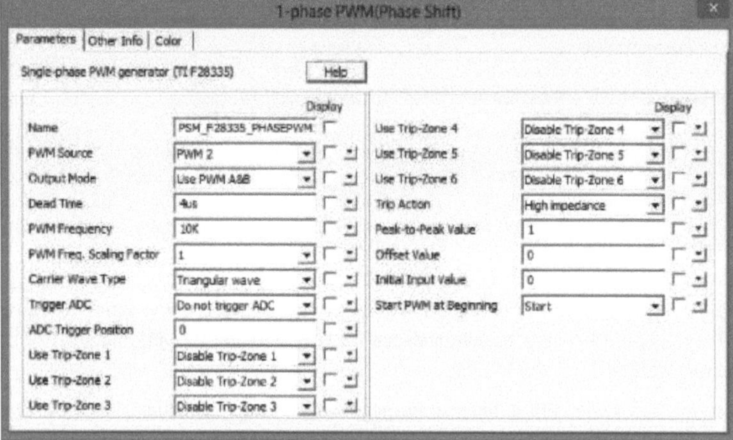

Source: PSIM.®

3.1.4 Two-phase PWM

The two-phase PWM block shown in Figure 25 has two inputs and two outputs, but unlike the single-phase generator, the outputs are not complementary. This generator outputs two signals of equal frequencies, generated by the same carrier. This generator does not have the option of using output A or B, i.e. both outputs are always activated. It also doesn't have the option of choosing the dead time. This PWM generator has six modes of use, differing in the type of carrier, triangular or sawtooth, and the mode of comparison with the carrier.

Figure 25- Two-phase PWM block.

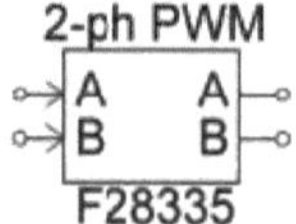

Source: Author's own production.

In the settings window for this block, shown in Figure 26, the following parameters can be changed:

- PWM *Source:* Definition of the corresponding ports to be used by the DSC. Channels PWM1 to PWM6 are found, and each channel uses two GPIO ports. For PWM1, the GPIO0 and GPIO1 ports are used, for PWM2, the GPIO2 and GPIO3 ports are used, following this pattern up to PWM6. It should be noted that six of these modules can be used in one application.
- *Mode Type*: Definition of the operating mode of outputs A and B, as explained later.

- PWM *Frequency*: Definition of the frequency of the PWM block, in hertz, this being the sampling frequency of the A/D Converter. When you open the settings window, this space will already be filled with 10 kHz.

- PWM *Frequency Scaling Factor*: You must choose between factors 1, 2 or 3. This scaling factor multiplies the PWM frequency, determining the frequency of the PWM block's output signal.

- ADC *Trigger*: You must choose whether the PWM block will trigger the A/D Converter and which group of the converter will be triggered, group A, group B, or both groups.

- ADC *Trigger Position*: You must choose whether the trigger will be made on an A/D converter, and whether it will be made at the beginning or in the middle of the carrier.

- *Use Trip-zones*: Definition of how *Trip-zones* 1 to 6 will be used, between disabling them, or enabling them in *One-shot* or *Cycle-by-Cycle* mode.

- *Trip Action*: Definition of whether the PWM generator will respond to the action of a *Trip-zone*, with the option of high impedance for both outputs, for only output A or only B, or no action at all.

Figure 26 - Configurations of the two-phase PWM block.

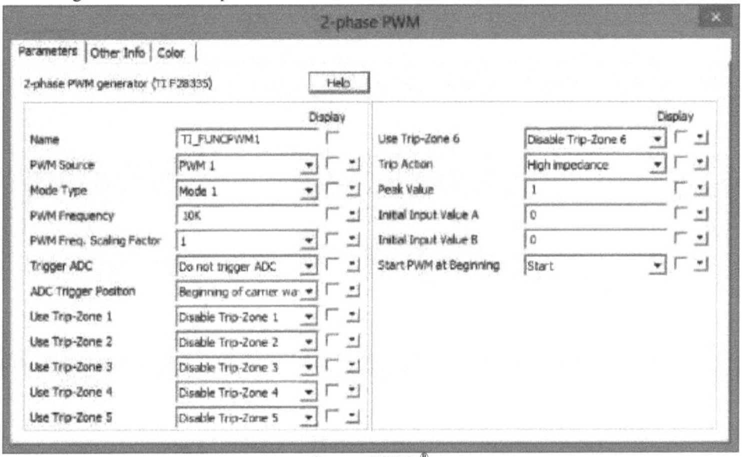

Source: PSIM.®

- *Peak-to-Peak Value:* Definition of the peak-to-peak value of the input signal.

- *Initial Input Value:* Definition of the initial value of inputs A and B. This value only affects the first few cycles of the output signal.

- *Start PWM at Beginning*: Selection between "*Start*", where the PWM generator will start activated, or "*Do not Start*", where the PWM generator will depend on a *Start* PWM module to work.

The two-phase PWM generator has six operating modes, which are explained below and can be seen in Figure 27.

1) Mode 1: Uses a sawtooth wave as the carrier. In this mode, while the input signals are greater than the carrier, the output signals remain at high logic level. This operating mode is

shown in Figure 27a.

2) Mode 2: Uses a sawtooth wave as the carrier. In this mode, the opposite of mode 1 occurs, i.e. while the input signals are higher than the carrier, the output signals remain at a low level, as shown in Figure 27b.

3) Mode 3: Uses a sawtooth wave as the carrier. In this mode, the generator inputs only change the AC output, with the CB output remaining unchanged at half the frequency and a cyclic ratio of 0.5. As can be seen in Figure 27c, the AC input determines the point at which the PWMxA output signal rises to the high level, while the B input determines the point at which the PWMxB output returns to the low logic level.

Figure 27 - Operating modes of the Two-Phase PWM block: (a)-(f) correspond to modes I to VI.

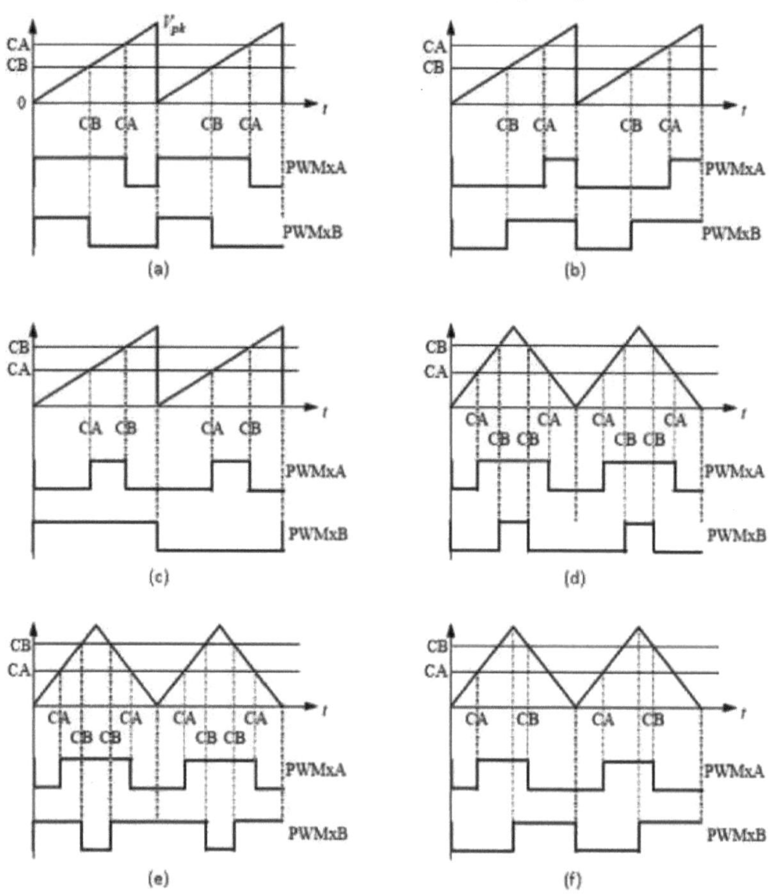

Source: PSIM User Manual (POWERSIM, 2010).

4) Mode 4: Uses a triangular wave as the carrier. In this mode, while the input signals are greater than the carrier, the output signals remain at a low logic level, as can be seen in Figure 27d.

5) Mode 5: Uses a triangular wave as the carrier. In this mode, the PWMxA output behaves in the same way as mode 4, but the PWMxB output behaves in the opposite way: as long as

the input signal is greater than the carrier, the output signal remains at a high level, as shown in Figure 27e.

6) Mode 6: Uses a triangular wave as the carrier. In this mode, the inputs only interfere with the PWMxA output, while the PWMxB output remains fixed with the same frequency as the carrier and a cyclic ratio equal to 0.5. The AC input determines the point at which the output signal goes from the low logic level to the high logic level, with the carrier's rising ramp as a reference, while the CB input determines the point at which the signal returns to the low level, with the carrier's falling ramp as a reference.

Modes 1, 2, 4 and 5 can be used when you have two reference signals and don't want the complementary PWM signals. Modes 3 and 6 can be used to obtain the difference between two constant signals.

3.1.5 Three-phase PWM

The three-phase PWM block consists of three single-phase modulators in just one block. This modulator has three reference inputs, one for each phase, called u, v and w, and six outputs, two for each input, as shown in Figure 28. The outputs corresponding to the u, v and w inputs are represented by the indices p and n, representing positive and negative signs respectively, indicating that they are complementary.

Figure 28 - Three-phase PWM block.

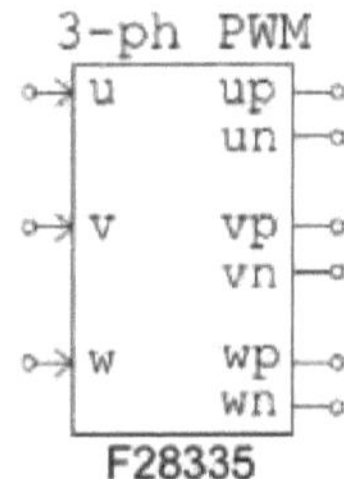

Source: Author's own production.

The following parameters are available for configuring the three-phase PWM modulator:

- PWM *Source:* Selection of the corresponding ports to be used by the DSC. They are divided into two channels, PWM123 and PWM456, and each channel uses six GPIO ports. For the PWM123 channel, ports GPIO0 to GPIO5 are used and for the PWM456 channel, ports GPIO6 to GPIO11 are used. Note that only two three-phase generators can be used in an application.
- *Dead Time*: Definition of the dead time between output signals, in seconds. This time is necessary, for example, so that two switches are not commanded to drive at the same instant of time. When you open the settings window, this space will already be filled with 4μs.
- PWM *Frequency*: Definition of the PWM frequency, in hertz, being the sampling frequency of the A/D Converter. When you open the settings window, this space will already be filled with 10 kHz.
- PWM *Frequency Scaling Factor*: You must choose between factors 1, 2 or 3. This scaling factor multiplies the PWM frequency, determining the frequency of the PWM block's output signal.
- *Carrier Wave Type*: Definition of the carrier wave type. You can choose between a

triangular or sawtooth waveform.

- ADC *Trigger*: You must choose whether the PWM block will trigger the A/D Converter and which group of the converter will be triggered, group A, group B, or both groups.
- ADC *Trigger Position:* You must choose whether to trigger on an A/D converter, and whether to trigger on group A, B or both groups of the converter.
- *Use Trip-zones*: You can choose how to use *Trip-zones* 1 to 6, whether to disable them or enable them in *One-shot* or *Cycle-by-Cycle* mode.
- *Trip Action*: You must choose how the PWM block will respond to the action of a *Trip-zone*. You can choose between high impedance for both outputs, for just the *p* output or just *n*, or no action at all.
- *Peak-to-Peak Value*: Definition of the peak-to-peak value of the input signal.
- *Offset Value*: Fill in the *offset* value of the input signal.
- *Initial Input Value*: This must be filled in with the initial values of the *u*, *v* and *w* inputs. This value only affects the first few cycles of the output signal.
- *Start PWM at Beginning*: You must select between "*Start*", in which case the PWM generator will start activated, or "*Do not Start*", in which case the PWM generator will depend on a *Start* PWM block to work.

Figure 29 shows the settings window for the three-phase PWM modulator.

Figure 29- Three-phase PWM modulator configurations.

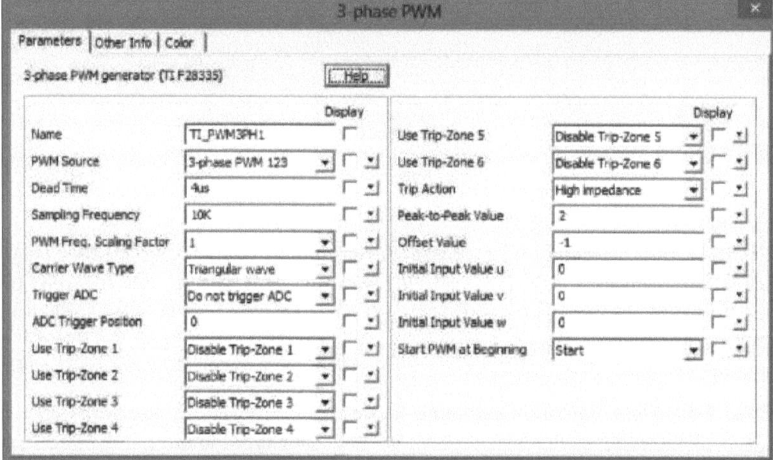

3.2. START PWM AND STOP PWM

The Start PWM and Stop PWM blocks, shown in Figure 30, have the function of activating or deactivating the PWM generators. Both components have only one input, and the block is activated if the input is at a high logic level (value greater than 0.5).

Figure 30 - Blocks (a) *Start* PWM and (b) *Stop* PWM.

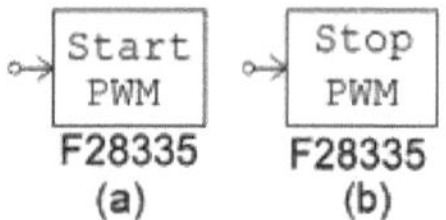

Source: Author's own production.

To configure both blocks, simply choose which PWM channel will be activated or deactivated. You can choose between the PWM1 to PWM6 channels for the one- or two-phase PWM blocks, or the PWM123 or PWM456 channels if the block is three-phase PWM, or Capture from 1 to 6, which can be used for both the Capture blocks, which will be shown next, and the Single PWM modulators. Figure 31 shows the block configuration windows.

Figure 31 - Configurations of the (a) *Start* and (b) *Stop* PWM blocks.

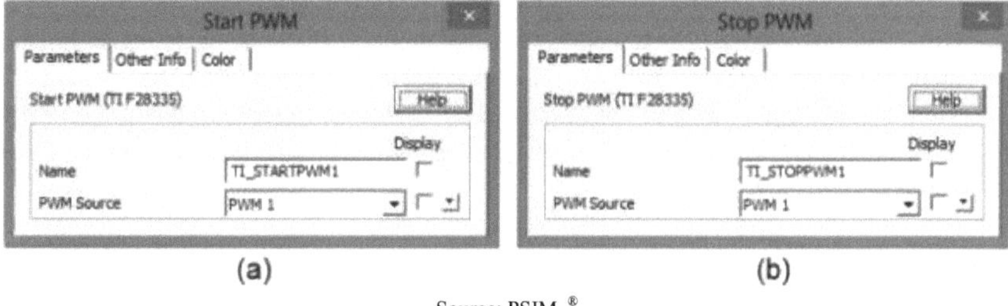

Source: PSIM.®

3.3. TRIP-ZONE AND TRIP-ZONE STATE

This block is used to interrupt PWM modulators. The Trip-Zone block has six inputs, each corresponding to a GPIO port. There are only six GPIO ports available for use, so only one Trip-Zone block can be used per application. The Trip-Zone action is activated with the input at a low logic level. As shown in Figure 32, this block does not have any output pins.

Figure 32 - *Trip-Zone* block.

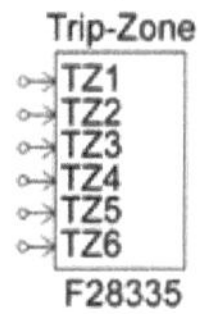

Source: Author's own production.

The available ports are GPIO12 to GPIO17, corresponding to *Trip-Zones* 1 to 6. A PWM modulator can be configured to be interrupted by more than one *Trip-Zone*. Figure 33 shows the *Trip-Zone* settings window.

Figure 33 - *Trip-Zone* block settings.

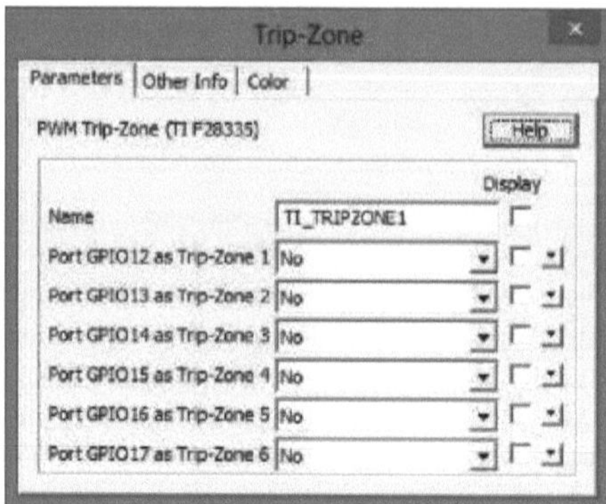

The Trip-Zone can operate in two modes, One-shot mode or Cycle-by-cycle mode. In One-shot mode, once the output of the PWM block has been interrupted, it must be restarted in order to function again. In Cycle-by-cycle mode, only the current cycle of the PWM modulator is affected by the interruption. The operating mode must be chosen in the PWM block settings.

The Trip-Zone status block has only one output, as shown in Figure 34, and its function is to indicate the operating mode of the Trip-Zone signal. This block is activated when the PWM modulator suffers an interruption, with its output at high logic level for One-shot mode and at low logic level for Cycle-by-cycle mode. The priority option for the Trip-Zone channels is not checked.

Figure 34 - *Trip-Zone State* block.

Source: Author's own production.

To configure this block, all you have to do is choose which PWM will be used to supervise it. You can therefore use one of these components for each PWM block. Its configuration window is shown in Figure 35.

Figure 35 - *Trip-Zone State* block settings.

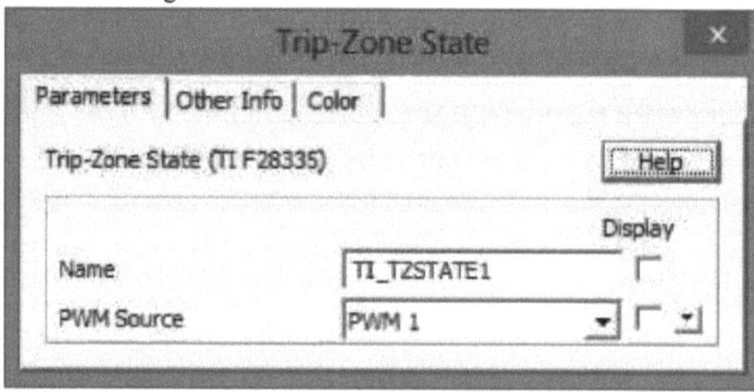

1.4. A/D CONVERTER

The A/D converter block has sixteen inputs, divided into two groups A and B. The voltage at each input must be between 0 and 3V in DC mode, or -1.5V to 1.5V in AC mode, but this mode is not possible with the DSC. Each input has a corresponding output, which can have a voltage gain, which can be used to compensate for the attenuations common to measurement circuits. The A/D converter can be seen in Figure 36.

Figure 36 - A/D converter block.

ADC

A0	D0
A1	D1
A2	D2
A3	D3
A4	D4
A5	D5
A6	D6
A7	D7
B0	D8
B1	D9
B2	D10
B3	D11
B4	D12
B5	D13
B6	D14
B7	D15

F28335

Source: Author's own production.

Figure 37 shows the A/D Converter's configuration parameters. The first parameter is the mode, where you can choose between Continuous, 8-channel Start/Stop or 16-channel Start/Stop. In Continuous mode, the converter continuously converts all inputs. In Start/Stop mode, the conversion is only carried out on request, for just one of the eight channel groups or for both groups, i.e. all 16 channels. This mode is used to trigger the PWM generators. For each channel used in the A/D converter, a mode must be set, between AC and DC, and the output gain.

Figure 37 - Configuring the A/D converter block.

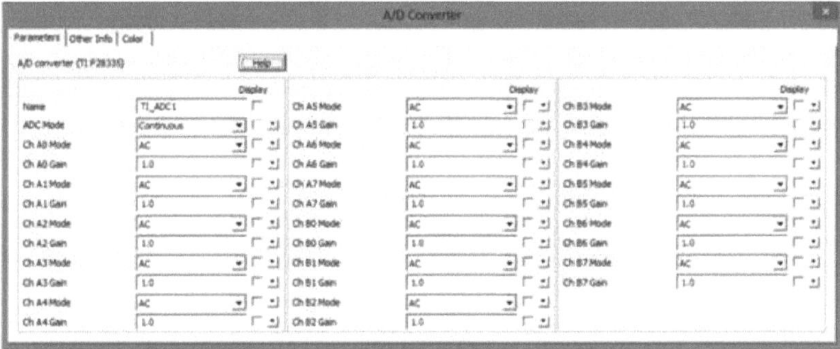

Source: PSIM .®

1.5. DIGITAL INPUTS AND OUTPUTS

The function of these blocks is to configure the digital inputs and outputs. Both have eight inputs and eight outputs. Figure 38 shows the digital input block.

Figura 38 - Digital Input Block.

DIN

D0	D0
D1	D1
D2	D2
D3	D3
D4	D4
D5	D5
D6	D6
D7	D7

F28335

Source: Author's own production.

In the Digital Inputs block, you need to configure which GPIO ports will be used and whether they will be used as an external interrupt. The settings window for this component is shown in Figure 39.

Figura 39 - Configuring the Digital Input block.

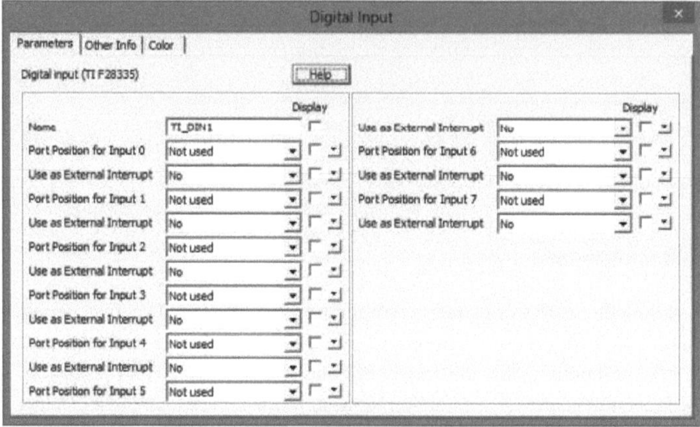

Source: PSIM .®

Any of the eighty-eight available ports can be configured as a digital input or output, and the GPIO0 to GPIO63 pins can be used for external interrupts. Therefore, several digital input and output blocks can be used in a simulation, but care must be taken not to use the same port in more than one block.

As for the Digital Output block, shown in Figure 40, only the GIPO gates to be used in the application must be configured. It should be noted that both the Digital Input and Digital Output blocks must be inserted into the application in such a way that they do not interfere with the voltage values, i.e. the input voltage of the block is the same as the output voltage.

Figura 40 - Digital Output Block.

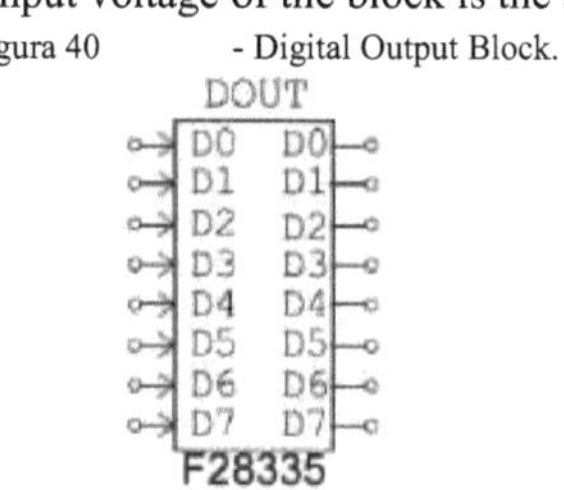

Source: Author's own production.

The configurations of the Digital Outputs block can be seen in Figure 41.

Figure 41 - Configurations for the Digital Outputs block.

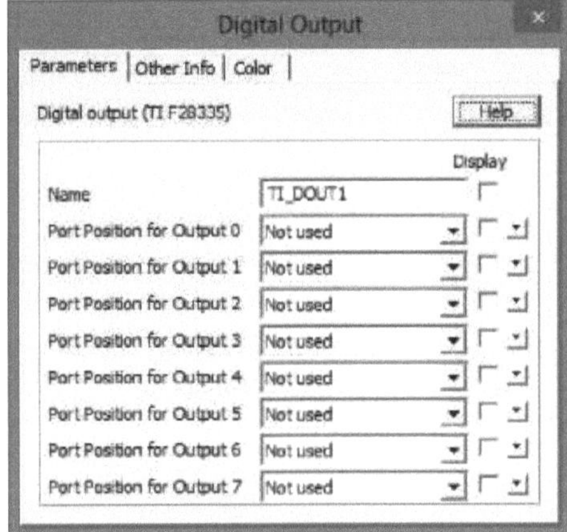

Source: PSIM.®

3.6. CAPTURE AND CAPTURE STATE

The Capture block, shown in Figure 42, is capable of generating interrupts from the input signal.

Figure 42 - Capture block.

Source: Author's own production.

In the configurations of this block, we find:

- *Capture Source:* Definition of the corresponding port to use. You can see that the ports available for this component are the same as for the single PWM generator.
- *Event Filter Prescale*: Definition of the value that will divide the input signal.
- *Timer* Mode: The time counting mode must be chosen between absolute or differential mode.

Figure 43 shows the settings window for this component.

Figure 43 - Capture block settings.

Capture (single counter) (TI F28335)	
Name	TI_CAPSINGLE1
Capture Source	Capture 1 (GPIO5)
Event Filter Prescale	No prescale
Timer Mode	Absolute time

Source: PSIM.®

The Capture Status component, shown in Figure 44, has a high logic level at its output when the capture signal is activated, and a low logic level when the capture signal is deactivated.

Figure 44 - Capture status block.

Source: Author's own production.

To configure it, all you have to do is select the source of the capture signal, i.e. which port is being used, as shown in Figure 45.

Figure 45 - Settings for the status capture block.

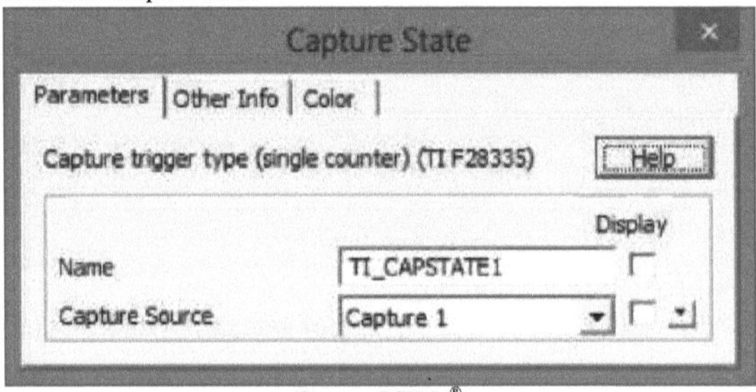

Source: PSIM .®

3.7. ENCODER AND ENCODER STATE

This block configures an Encoder input. As shown in Figure 46, this component has four inputs and one output. The A and B inputs should have square waves offset by 90°. The Z input indicates that the Encoder has turned round and should have a square signal as an input. The Strobe input can be used as an interrupt source, as can the Z input.

Figure 46 - Encoder block.

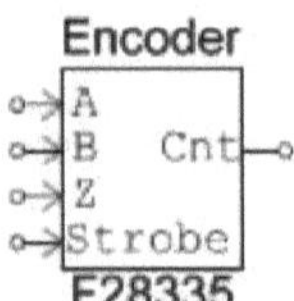

Source: Author's own production.

This element counts the number of pulses from the A and B inputs, with the Z and Strobe inputs as the counting reference. However, you can choose not to use the last two inputs, causing the Encoder to count infinitely.

To configure this component, we find the following items:

- *Encoder Source:* You must choose which GPIO ports will be used. You can use up to two *Encoders* per application, configuring them as *Encoder1*, with GPIO ports 20 and 21 or 50 and 51, or as *Encoder2*, with GPIO ports 24 and 25, where one port corresponds to the *Cnt* output and the other to the *Encoder State* block output.
- *Use Z Signal*: You can choose to use the Z input or not.
- *Use Strobe Signal*: You can choose to use the Strobe input or not.
- *Counting Direction*: Definition of the counting direction, with *Forward* being the ascending order and *Reverse* the descending order.
- *Encoder Resolution*: Definition of the *Encoder*'s resolution, being the number of pulses

per revolution.
These configurations can be seen in Figure 47.

Figure 47 - *Encoder* block settings.

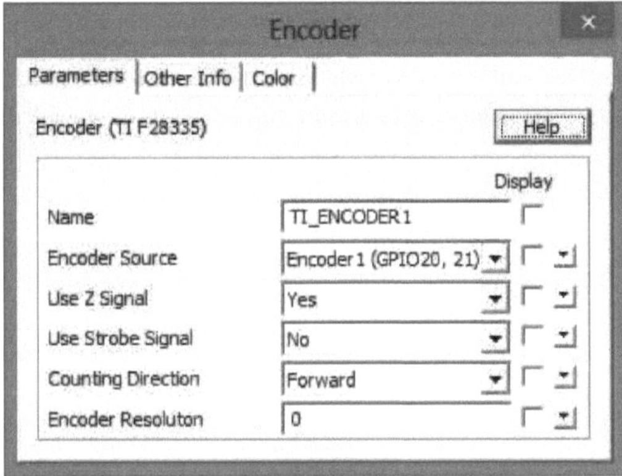

Source: PSIM .®

The Encoder State block shown in Figure 48 has the function of indicating the source of the interruption. Its output shows a high logic level when the Z signal causes the interruption, and a low logic level when the Strobe signal generates the interruption.

Figure 48 - *Encoder State* block.

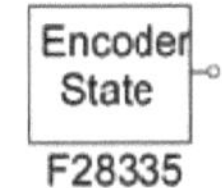

Source: Author's own production.

To configure it, simply choose which ports will be used. Remember to use the same ports that were configured for the *Encoder*. Figure 49 shows the *Encoder State* configuration window.

Figure 49 - *Encoder State* block configurations.

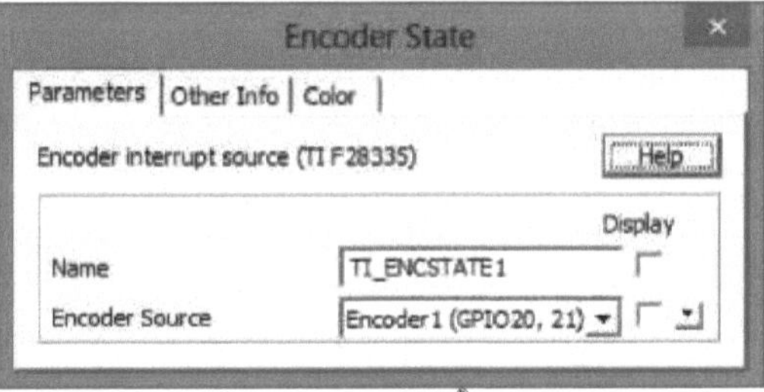

Source: PSIM .®

3.8. UP/DOWN COUNTER

This Counter is basically a simplified version of the *Encoder*. It has just two inputs, *Clk* and *Dir*, and the *Cnt* output, as seen in Figure 50. With each pulse on the *Clk* input, the value of the *Cnt* output is incremented or decremented by one. The *Dir* input has the function of determining the counting direction. If the value of the Dir=1 input, the count is increasing, and if the value of Dir=0, the count is decreasing.

Figure 50 - *Up/Dowm* Counter block.

Source: Author's own production.

To configure this Counter, simply choose which pin will be used. You can see that the GPIO ports available are the same as those used for the *Encoder*, but the counter only uses one port. As with the *Encoder*, only two Counters can be used in an application. The *Up/Dowm* Counter settings window is shown in Figure 51.

Figure 51 - Settings for the *Up/Dowm* Counter block.

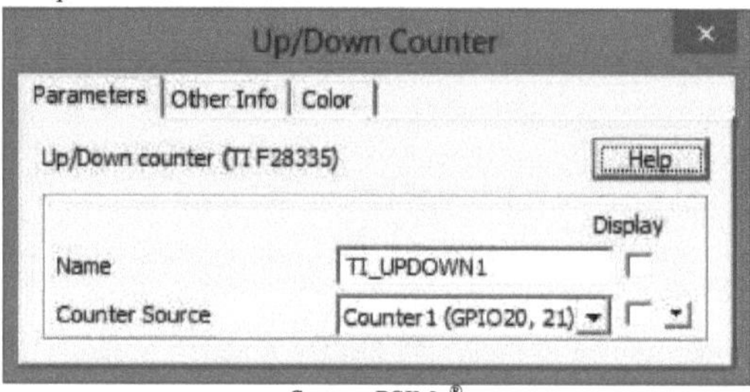

Source: PSIM .®

3.9. DSP CLOCK

The *SimCoder* tool has this component, capable of configuring an external *clock* source. The component is shown in Figure 52.

Figure 52 - DSP *Clock* block.

Source: Author's own production.

You can set the frequency of the external *clock* by entering an integer up to 30MHz, if you need to use it. You can also set the speed of the DSP, also in MHz, which must be an integer and a multiple of the value set for the external *clock,* with a maximum of 150MHz. Figure 53 shows the settings window for this component.

Figure 53 - DSP *Clock* block settings.

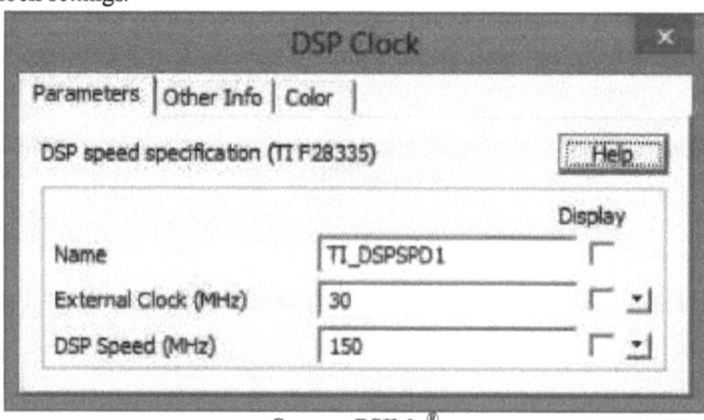

Source: PSIM .®

3.10. HARDWARE CONFIGURATION

Using this component, you can configure all the ports that will be used in the application. You can check which ports can be used for which functions. It is a very useful tool to avoid configuring two components with the same GPIO port. However, even using this component, it is necessary to configure each element individually. Figure 54 shows the Hardware Configuration component.

Figura 54 - Hardware Configuration Block.

Source: Author's own production.

After configuring it, you must lock this component using the Lock button, as shown in Figure 55. To unlock it, simply press it again.

Figura 55 - Settings from the Hardware Configuration block.

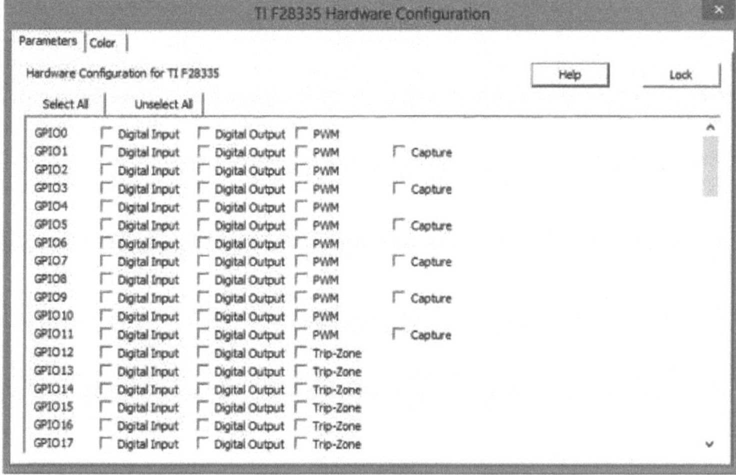

Source: PSIM .®

3.11. OTHER COMPONENTS

In addition to the specific components for the DSC TI F28335, you can use other components found in PSIM. The components can be found under Elements, identified by the symbol shown in Figure 56. Elements include mathematical operation blocks, logic elements, digital control elements and more.

Figura 56 - Symbol for elements that can be used to generate code.

Source: Author's own production.

The PSIM software differentiates between power circuits and control circuits. Power circuits have red connections, while control circuits have green connections. The SimCoder components and other elements identified by the symbol above are control elements, so they cannot be connected to components such as resistors, transistors, operational amplifiers, among others.

For a SimCoder element, such as an A/D converter or an analogue input, to read a voltage or current value in a simulated circuit, it is necessary to use a voltage or current sensor, shown in Figure 57. In this way, the power value is transformed into a control signal.

Figura 57 - Sensor blocks for: (a) voltage and (b) current.

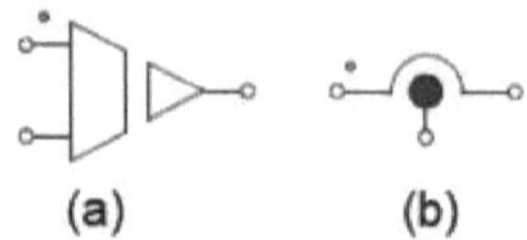

Source: Author's own production.

To insert an output from a SimCoder component, such as a PWM generator, as an input into a simulated circuit, it is necessary to use an On-Off controller, which can be seen in Figure 58. In this way, the control signal is transformed into a voltage value that can be used to drive a transistor, for example, in a power circuit.

Figura 58 - On-Off Controller.

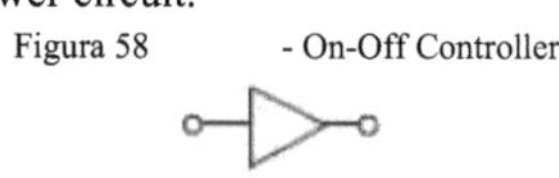

Source: Author's own production.

It may be necessary to use an input with a sinusoidal signal in simulations, for example when modulating the control of a DC-AC converter. However, SimCoder's PWM generators do not accept a sinusoidal voltage source as an input. A possible solution is to use an external sinusoidal source and an A/D converter. However, you can use some PSIM elements that are normally accepted for code generation. Figure 59 shows how the sinusoidal signal is generated.

Figura 59 - Sinusoidal signal used for the simulations.

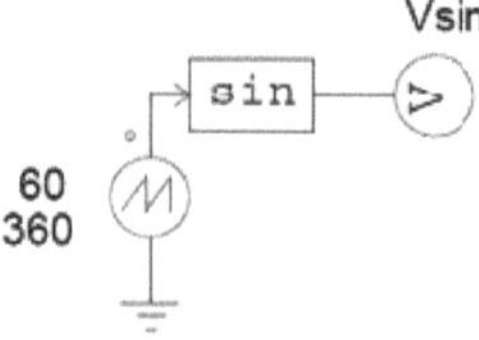

Source: Author's own production.

To generate a sine wave, simply use a sawtooth wave generator with an amplitude of 360 and a sine function. This produces a sine wave of unit amplitude, with the same frequency as the sawtooth wave, which can be used as a reference for a PWM generator. The sine function is found in Elements>Control>Computational Blocks and the sawtooth generator in Elements>Sources>Voltage. Figure 60 shows the waveform of the signal obtained.

Figure 60 - Sine signal obtained using a sawtooth wave and a sine function.

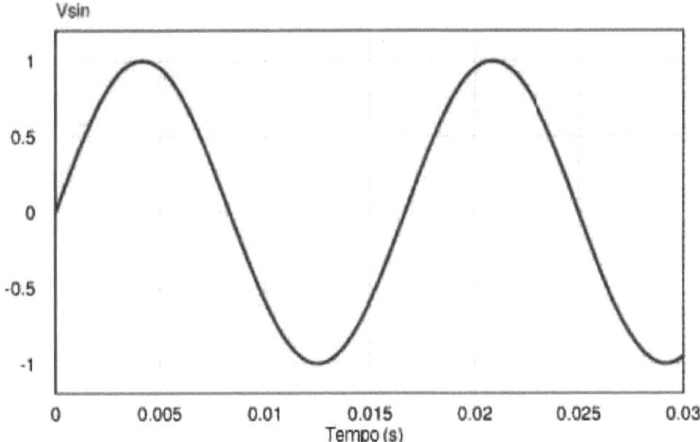

Source: Author's own production.

CHAPTER 4

4. GENERATED CODE AND CODE COMPOSER STUDIO

After studying the elements present in SimCoder, it is possible to carry out simulations using these components and then generate code in C language for application in Code Composer Studio.

4.1. STUDY OF THE GENERATED CODE

Figure 61 shows a simulation circuit for testing some of the elements. An A/D converter was used to obtain an analogue input. The value obtained from the A/D converter is used as an input for a single-phase PWM generator. A digital input, a Start PWM and a Stop PWM were also used to switch the PWM generator on and off. Finally, a digital output was used just to show the operating status of the PWM generator.

Figure 61 - Simulation circuit for testing some *SimCoder* elements.

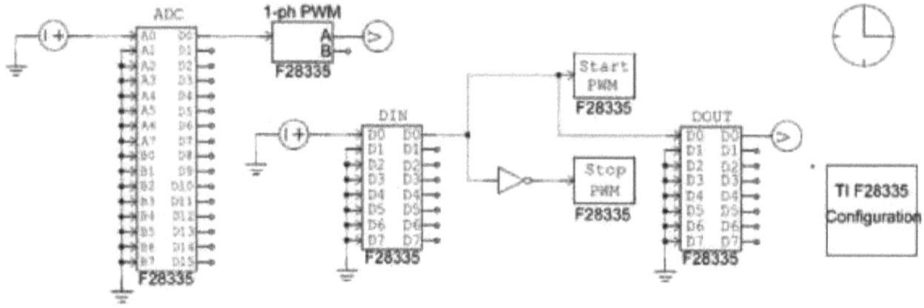

Source: Author's own production.

The A/D converter was configured for continuous reading, in DC mode and with unity gain. The PWM block and the PWM *Start and Stop* blocks have all been configured for the PWM1 channel, which uses the GPIO0 and GPIO1 ports. In addition, the PWM block has been configured to have a 0 to 3V input and not to start switched on. The GPIO5 and GPIO6 ports were used for the digital input and output respectively, and the digital input was not used as an external interrupt.

The PWM modulator was set to a peak-to-peak value of 3V, an offset value of 0, a dead time of 0.5LIS and a frequency of 20kHz. Figure 62 shows the settings window for the PWM modulator used.

Figure 62 - Configured parameters of the PWM block used for testing.

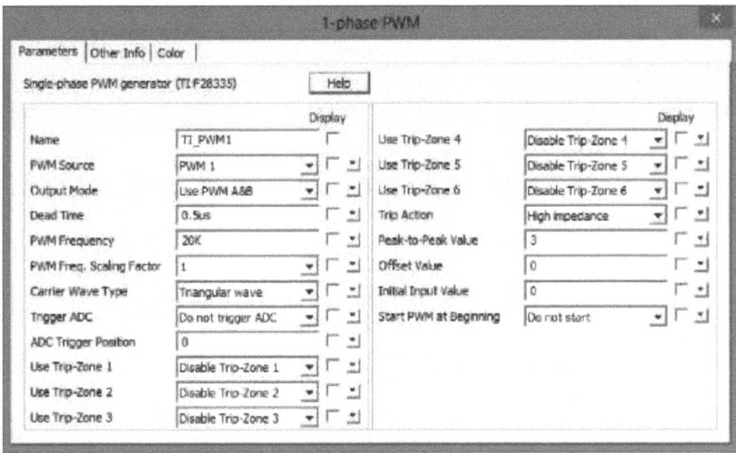

Source: PSIM .®

Finally, all the unused inputs of the components were properly grounded. In the hardware settings block, the ports used were marked and then locked. The TI F28335 option was also selected in the Hardware Target of the simulation control. To finally generate the code, the application was simulated and Generate Code was selected in Simulate. The C code obtained can be divided into parts to aid understanding, as shown below.

The first part of the code aims to initialise the libraries used and present the program's functions. You can see that one of the libraries is PS_bios.h, generated from the PSIM simulation.

```
/*********************************************************************
// ThiscodeiscreatedbySimCoderVersion 2.0 for F28335 Hardware Target
//
// SimCoderis copyright byPowersim Inc., 2009
//
// Date: May 07, 2014 17:27:12
*********************************************************************/
#include     <math.h>
#include     "PS_bios.h"
TypedeffloatDefaultType;
#define      GetCurTime() PS_GetSysTimer()

interruptvoidTask();
voidTask_1();
```

The part of the code below represents a function responsible for reading the analogue input and determining the PWM duty cycle. If there were any logic involving the value of the analogue input or the PWM duty cycle, it would be in this part of the code. This function does not need to be called because it is of the interrupt type, i.e. it is activated every sampling period. The PS_GetDcAdc function reads the value of the A/D converter's 0 input, as done in the PSIM simulation.

```
interruptvoidTask()
{
	DefaultType fTI_ADC1;
	PS_EnableIntr();

	fTI_ADC1 = PS_GetDcAdc(0);
	PS_SetPwm1Rate(fTI_ADC1);
	PS_ExitPwm1General();
}
```

This section is the function responsible for switching the PWM on and off. This function also checks the digital input and determines the value of the digital output. In the PS_StartPwm and PS_StopPwm, you can see that both are associated with PWM1, which will be initialised next.

```
voidTask_1()
{
	DefaultType fTI_DIN2, fNOT1;

	fTI_DIN2 = (PS_GetDigitInA() & ((Uint32)1 << 5)) ? 1 : 0;
	if (fTI_DIN2 > 0)
	{
		PS_StartPwm(1);
	}
	fNOT1 = !fTI_DIN2;
	if (fNOT1 != 0)
	{
		PS_StopPwm(1);
	}
	(fTI_DIN2 == 0) ?PS_ClearDigitOutBitA((Uint32)1 << 6) :
PS_SetDigitOutBitA((Uint32)1 << 6);
}
```

This last function initialises all the program elements, i.e. the PWM, analogue input, digital input and output. If there were other elements in the simulation, they would all be configured in this section. Comparisons can be made between the generated code and the configurations of the elements in the PSIM simulation. We can see in the PS_InitPwm function that the following are configured: the PWM number; the type of carrier wave, 1 for triangular wave and 0 for sawtooth; the PWM frequency, obtained by multiplying the sampling frequency by the scaling factor; the dead time and the output mode. The functions below show the PWM input settings, the A/D converter trigger and the A/D converter settings. Finally, you can see the initialisations of the digital input and output and the PWM Start and Stop blocks.

```
voidInitialize(void)
{
        PS_SysInit(30, 10);
        PS_StartStopPwmClock(0);
        PS_InitTimer(0, 0xffffffff);
        PS_InitPwm(1, 1, 20000*1, (0.5e-6)*1e6, PWM_TWO_OUT, 34968);     // pwnNo,
waveType, frequency, deadtime, outtype
        PS_SetPwmPeakOffset(1, 3, 0, 1.0/3);
        PS_SetPwmIntrType(1, ePwmNoAdc, 1, 0);
        PS_SetPwmVector(1, ePwmNoAdc, Task);
        PS_SetPwm1Rate(0);
        PS_StopPwm(1);

        PS_ResetAdcConvSeq();
        PS_SetAdcConvSeq(eAdcCascade, 0, 1.0);
        PS_AdcInit(0, !0);

        PS_InitDigitIn(5, 100);

        PS_InitDigitOut(6);

        PS_StartStopPwmClock(1);
}
```

Finally, you can see the main part of the code. First the function to initialise the program is called, then it enables the internal interrupt and finally it enters an infinite loop in which the Task_1 function, responsible for operating the PWM, is constantly called.

```
voidmain()
{
        Initialize();
        PS_EnableIntr();   // Enable Global interrupt INTM
        PS_EnableDbgm();
        for (;;) {
                Task_1();

        }
}
```

It's important to understand the generated code, because you can make small changes to the settings. However, when a change is made to the PSIM simulation® and new code is generated, Code Composer Studio allows the code to be updated automatically. This only happens if the code is in the same workspace as the one used.

4.2. DSC AND CODE COMPOSER STUDIO

The TMS320F28335 is a digital floating point signal controller from Texas Instruments. According to the user manual, the DSC has a maximum clock frequency of 150 MHz and a 6.67ns instruction cycle. It has a 16 to 32-bit external interface and a 12-bit A/D converter, with sixteen channels and a conversion time of 80 ns. It has six PWM channels, eighty-eight ports available for digital inputs and outputs and eight ports for external interrupts. The DSC has Harvard Bus architecture and 256kB of memory. Figure 63 shows the DSC experiment kit used.

Figure 63 - *Texas Instruments* TMS320F28335 experiment kit.

Source: TEXAS INSTRUMENTS, 2011.

Code Composer Studio (CCS) is the software developed for Texas Intruments (TI) DSCs, microcontrollers and microprocessors. The software includes a set of tools used to develop and debug code in the C language. It includes compilers for each of TI's device families, a source code editor, a project building environment, simulators and many other features.

The PSIM software, using the SimCoder tool, is able to generate code ready for implementation in the DSC. SimCoder generates code for CCS version 3.3. In this work, version 5.1 of this software will be used, shown in Figure 64.

When the code is generated in PSIM, several files are created simultaneously:

- .c" file: this file will have the same name as the PSIM file and will be the simulation's main code.
- ".pjt" file: CCS project file and also has the same name as the PSIM file.
- PS_bios.h: Header file for SimCoder's F28335 libraries.

Figure 64 - Code Composer Studio version 5.1.

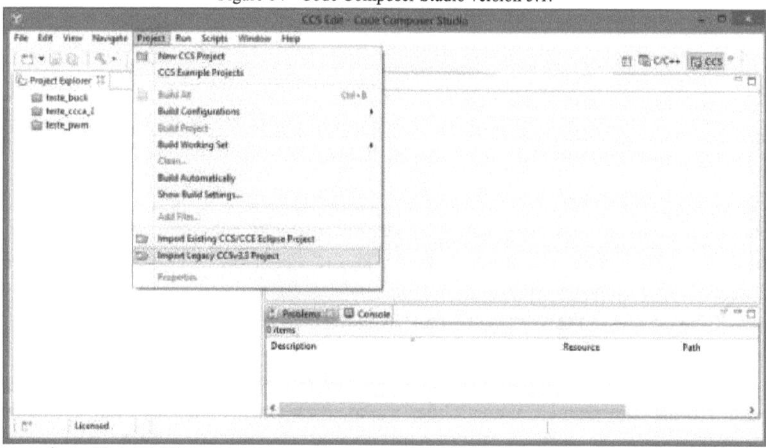

Source: Author's own production.

- PsBiosFlash.lib: *SimCoder*'s F28335 library.

- PsBiosRam.lib: *SimCoder*'s F28335 library.
- C28x_FPU_FastRTS_beta1.lib: *Texas Instruments* library for floating-point DSC.
- Password.asm: file to specify the DSC password code.
- DSP2833x_Headers_nonBIOS.cmd: Peripheral register access file.
- F28335_FLASH_Lnk.cmd: *Flash* memory access file.
- F28335_FLASH_RAM_Lnk.cmd: *Flash Ram* memory access file.
- F28335_RAM_Lnk.cmd: *Ram* memory access file.

All the files shown are part of the project generated for *Code Composer Stutio* v3.3. To open this project in version 5.1 or later of this *software*, select *Project/Import Legacy* CCSv3.3 Project, as shown in Figure 64. Figure 65 shows the window for importing the project.

Figure 65 - Window for importing a project from CCS version 3.3.

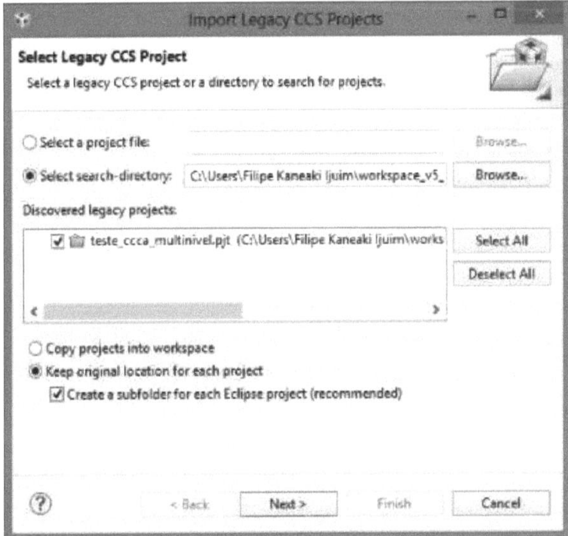

Source: Author's own production.

Once you've selected the project folder, simply click Next and then Finish to create the project for Code Composer Studio v5.1.

By selecting the file "teste_ccca_multinivel.c", you can see the code that was generated by SimCoder, as seen in Figure 66. Using this code, you can compile the project without having to make any changes.

Figure 66 - Visualisation of the code generated using CCSv5.1

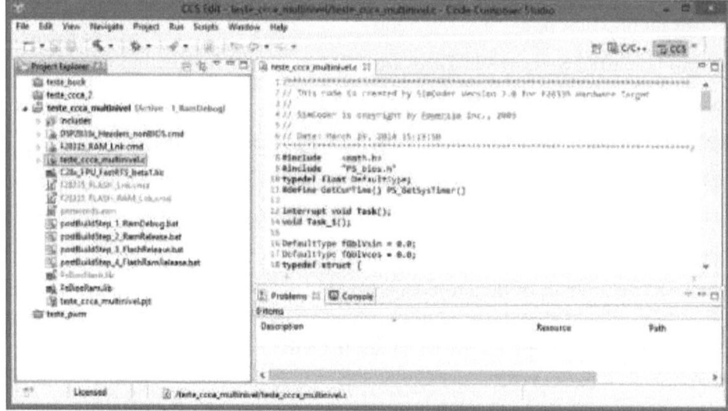

Source: Author's own production.

To make the process easier, we suggest always saving the PSIM simulation on the desktop used in CCS. If any changes are made to the PSIM simulation, the generated code will be automatically updated in CCS, preventing the whole import process from having to be done again.

CHAPTER 5

5. SIMCODER APPLICATIONS IN POWER ELECTRONICS CIRCUITS

After studying static converters, it is possible to carry out simulations in PSIM using the elements that make it possible to generate C code.

5.1 BUCK CONVERTER

To simulate the Buck converter, we used the elements in Figure 67, which show an A/D converter, a single-phase PWM modulator, a digital input and output, a Start PWM and a Stop PWM.

Figure 67 - Simulation and DSC settings for the simulated open-loop Buck converter.

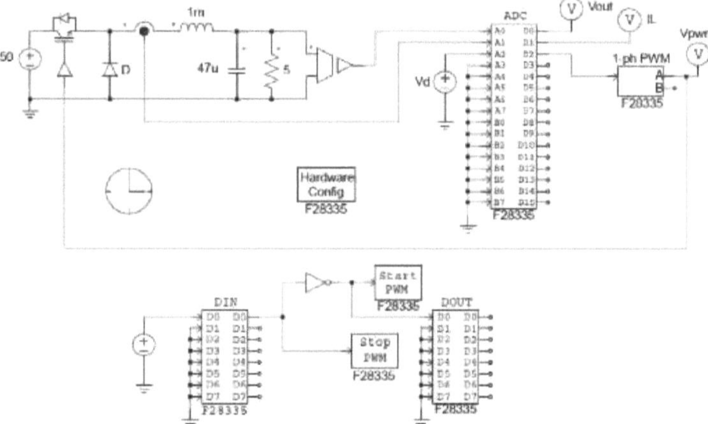

The functions of the A/D converter in the circuit are to obtain the values of the current and voltage sensors, and to obtain the reference value for the PWM block's cyclic ratio from an analogue signal. The PWM modulator has the function of controlling the converter switch. Note that the voltage and current sensors were determined with a gain of 1/64 and 1/4, respectively, so that the A/D converter's input value is below 3.0 V, i.e. within the voltage range allowed for the converter's DC mode. Figure 68 shows part of the A/D Converter settings window.

Figure 68 - Settings window for the A/D converter used in the simulation of the open-loop Buck converter Buck converter simulation.

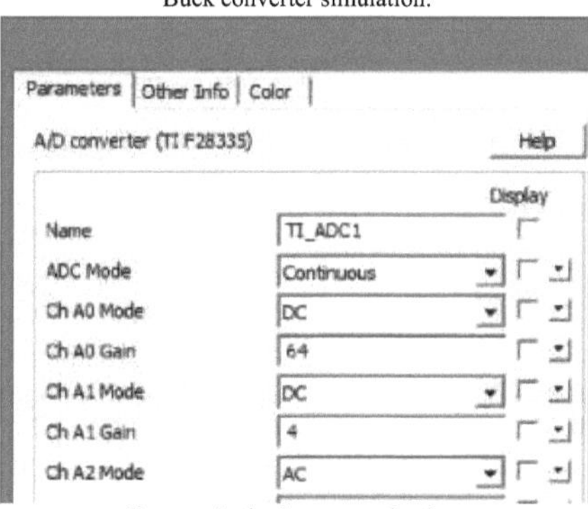

Source: Author's own production.

The A/D converter was configured with a gain of 64 and 4 for channels A0 and A1, respectively, to compensate for the reduction in voltage and current sensors. Unit gain was used for channel A2 and all the channels used were in DC mode. The PWM block was configured as shown in Figure 69.

The digital input blocks, Start and Stop PWM, constitute a logic for activating and deactivating the PWM block, via a button that can be connected to the DSC. The digital output simply indicates whether the PWM block is switched on or not, and can be visualised via an LED.

Figure 69 - Configurations of the PWM block used for the Buck converter.

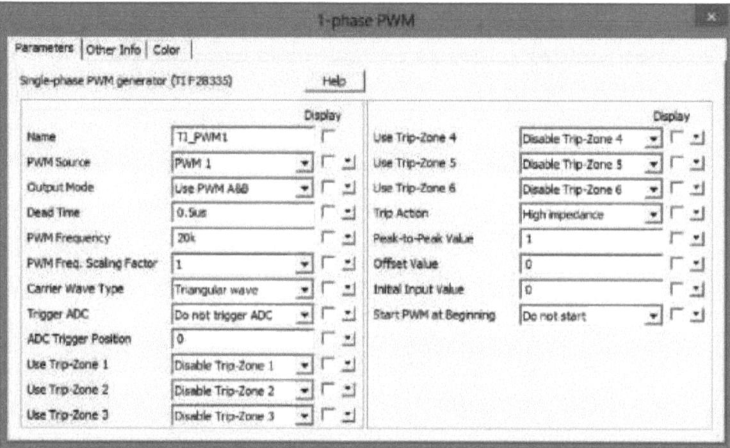

Source: Author's own production.

From the simulated circuit, we obtain the waveforms of the Buck converter in open loop, as shown in Figure 70.

Figure 70 - Buck converter output waveforms in open loop.

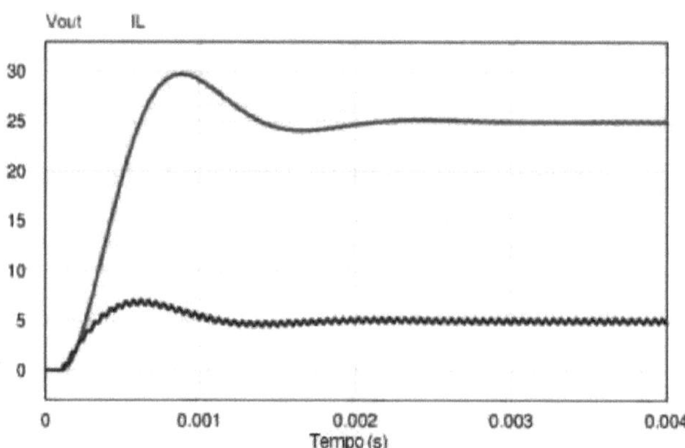

Source: Author's own production.

Based on the simulation carried out, code can be generated in C language for application in the Texas Instruments DSC, resulting in the code below.

```
/*****************************************************************************
// This code is created by SimCoder Version 2.0 for F28335 Hardware Target
//
// SimCoder is copyright by Powersim Inc., 2009
//
// Date: July 30, 2014 11:17:41
*****************************************************************************/
#include     <math.h>
#include     "PS_bios.h"
typedef float DefaultType;
#define      GetCurTime() PS_GetSysTimer()

interrupt void Task();
void Task_1();

DefaultType  fGblVadc2 = 0.0;
DefaultType  fGblVadc1 = 0.0;


interrupt void Task()
{
	DefaultType fTI_ADC1_2;
	PS_EnableIntr();

	fTI_ADC1_2 = PS_GetDcAdc(2);
	PS_SetPwm1Rate(fTI_ADC1_2);
	PS_ExitPwm1General();
}

void Task_1()
{
	DefaultType fTI_DIN1, fNOT1, fTI_ADC1_1, fTI_ADC1;

	fTI_DIN1 = (PS_GetDigitInA() & ((Uint32)1 << 10)) ? 1 : 0;
	fNOT1 = !fTI_DIN1;
	if (fNOT1 > 0)
	{
		PS_StartPwm(1);
	}
	if (fTI_DIN1 != 0)
	{
		PS_StopPwm(1);
	}
	fTI_ADC1_1 = PS_GetDcAdc(1);
#ifdef _DEBUG
	fGblVadc2 = fTI_ADC1_1;
#endif
	fTI_ADC1 = PS_GetDcAdc(0);
#ifdef _DEBUG
	fGblVadc1 = fTI_ADC1;
```

```
#endif
	(fNOT1 == 0) ? PS_ClearDigitOutBitA((Uint32)1 << 11) :
PS_SetDigitOutBitA((Uint32)1 << 11);
}


void Initialize(void)
{
	PS_SysInit(30, 10);
	PS_StartStopPwmClock(0);
	PS_InitTimer(0, 0xffffffff);
	PS_InitPwm(1, 1, 10000*1, (4e-6)*1e6, PWM_POSI_ONLY, 157);	// pwnNo,
waveType, frequency, deadtime, outtype
	PS_SetPwmPeakOffset(1, 1, 0, 1.0/1);
	PS_SetPwmIntrType(1, ePwmNoAdc, 1, 0);
	PS_SetPwmVector(1, ePwmNoAdc, Task);
	PS_SetPwm1Rate(0);
	PS_StopPwm(1);

	PS_ResetAdcConvSeq();
	PS_SetAdcConvSeq(eAdcCascade, 0, 4);
	PS_SetAdcConvSeq(eAdcCascade, 1, 4);
	PS_SetAdcConvSeq(eAdcCascade, 2, 1.0);
	PS_AdcInit(0, !0);

	PS_InitDigitIn(10, 100);

	PS_InitDigitOut(11);

	PS_StartStopPwmClock(1);
}

void main()
{
	Initialize();
	PS_EnableIntr();	// Enable Global interrupt INTM
	PS_EnableDbgm();
	for (;;) {
		Task_1();
	}
}
```

The simulation above was done in open loop. However, with the model obtained in the previous section, a controller can be designed with the aim of obtaining some desired characteristic, or eliminating some undesired characteristic. Figure 71 shows the Buck converter in closed loop, where the PI controller has a gain of 0.03 and a time constant of 0.0001s.

Figure 71 - Closed-loop buck converter with PI controller.

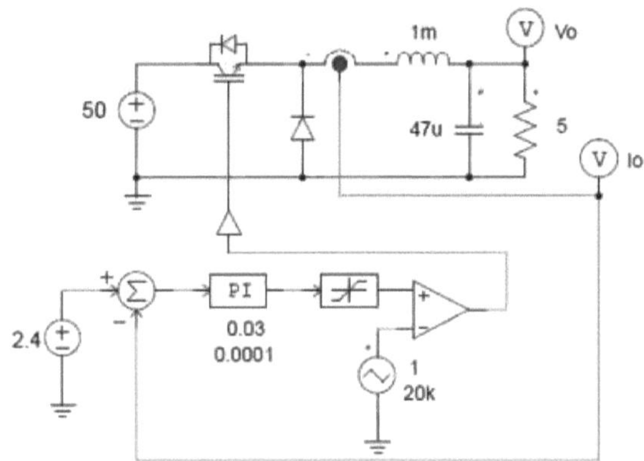

Source: Author's own production.

The voltage and current waveforms are shown in Figure 72.

Figure 72 - Buck converter output waveforms in closed loop with PI controller.

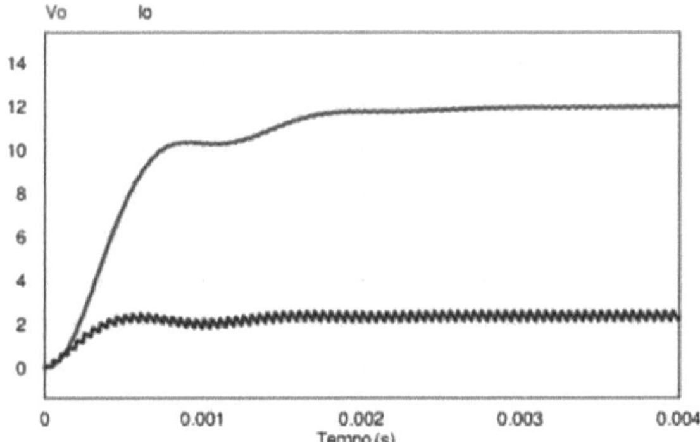

Source: Author's own production.

The PI controller shown in Figure 71 is represented in the frequency domain, so it must first be converted to the discrete (z) domain in order to be used by the SimCoder tool to generate the code in C language. One way of doing this is to use the s2z Converter tool in PSIM® , found in the Utilities menu. This tool is used to obtain the discretised values of the controller. Figure 73 shows the tool, displaying the values found.

Figure 73 - PSIM's s2z Converter tool .®

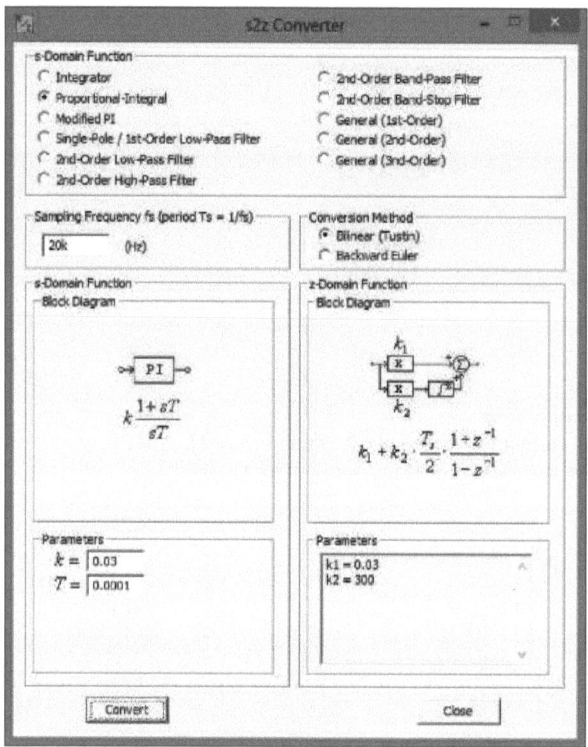

The s2z Converter tool already shows the equivalent circuit for the PI controller, giving the values for the proportional gain and integral gain. With the values obtained, the controller can be simulated together with the SimCoder elements, making it possible to generate the code in C language. Figure 74 shows the closed-loop Buck converter with the PI controller obtained.

Figure 74 - Buck converter with discretised PI controller.

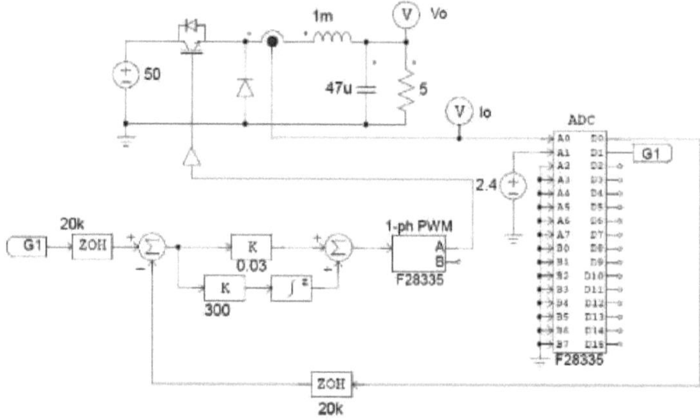

Source: Author's own production.

It should be noted that since we are no longer working with continuous time, zero-order hold (ZOH) must be used after the analogue readings. Figure 75 shows the Buck converter output waveforms using the discretised controller obtained. It can be seen that this waveform, as with the analogue controller, does not have an overshoot value and has a close settling time.

Figure 75 - Buck converter output waveforms with discretised controller.

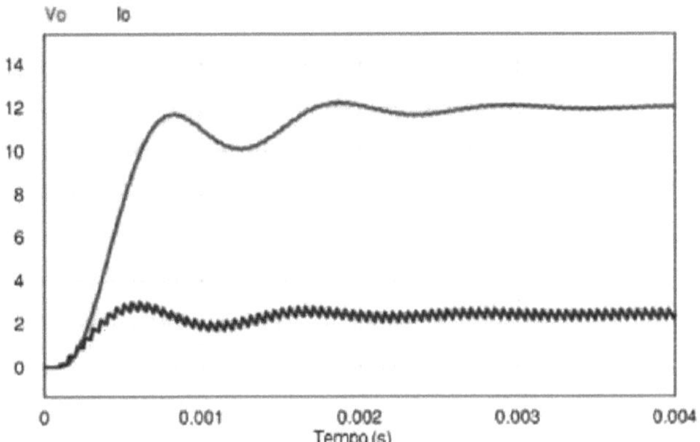

Source: Author's own production.

The part of the code corresponding to the simulated controller is shown below.

```
interrupt void Task()
{
	DefaultType fTI_ADC1_1, fZOH2, fTI_ADC1, fZOH1, fSUM1, fP1, fP2, fB5,
fSUMP1;
	PS_EnableIntr();

	fTI_ADC1_1 = PS_GetDcAdc(1);
	fTI_ADC1 = PS_GetDcAdc(0);
	fZOH2 = fTI_ADC1_1;
	fZOH1 = fTI_ADC1;
	fSUM1 = fZOH2 - fZOH1;
	fP1 = fSUM1 * 0.03;
	fP2 = fSUM1 * 300;
	{
		static DefaultType out_A = 0, in_A = 0.0;
		fB5 = out_A + 0.5/20000 * (fP2 + in_A);
		out_A = fB5;in_A = fP2;
	}
	fSUMP1 = fP1 + fB5;
	PS_SetPwm1Rate(fSUMP1);
	PS_ExitPwm1General();
}
```

Another option is to discretise the PI controller using the MATLAB software® . The controller model is found in its settings by pressing the Help button. The PI controller has the following transfer function in the *s* domain:

$$G(s) = k * \frac{1 + sT}{sT} \qquad (11)$$

Where *k* is the gain and *T* is the time constant of the PI controller. Applying the controller values and using MATLAB software® , the transfer function in the *z* domain was obtained using the c2d function, with a sampling time of 0.05ms and using the Tustin method.

$$G(z) = \frac{0{,}0375z + 0{,}0225}{z - 1} \qquad (12)$$

Figure 76 shows the circuit using the PSIM *z-domain* transfer function block• with the parameters of the transfer function obtained in MATLAB .•

Figure 76 - Buck converter with controller in the form of a z-domain transfer function.

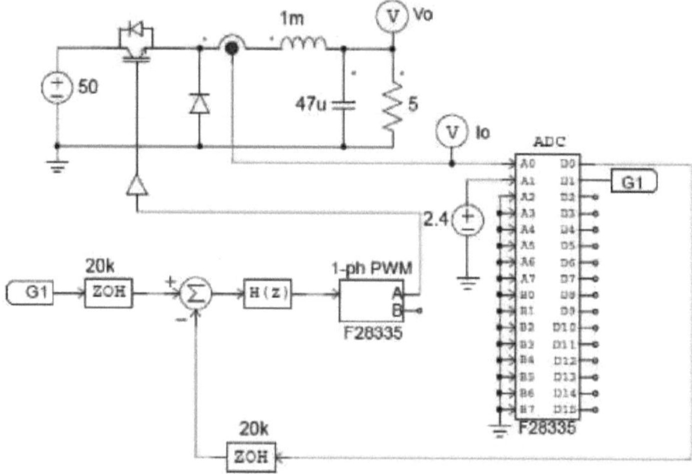

Source: Author's own production.

The output waveforms of the closed-loop converter with the controller in the form of the z-domain transfer function obtained are the same as the waveforms found with the discretised controller, as shown in Figure 77.

Figure 77 - Buck converter output waveforms with z-domain transfer function controller domain.

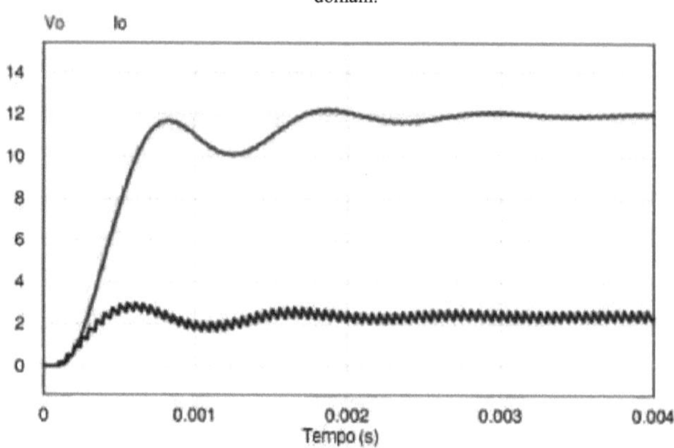

Source: Author's own production.

From the simulation, the code for the DSC can be generated. The table below shows part of this code, corresponding to the simulated controller.

```
interrupt void Task()
{
	DefaultType fTI_ADC1_1, fZOH2, fTI_ADC1, fZOH1, fSUM1, fTF_D2;
	PS_EnableIntr();

	fTI_ADC1_1 = PS_GetDcAdc(1);
	fTI_ADC1 = PS_GetDcAdc(0);
	fZOH2 = fTI_ADC1_1;
	fZOH1 = fTI_ADC1;
	fSUM1 = fZOH2 - fZOH1;
	{
		static DefaultType fIn = 0.0;
		static DefaultType fOut = 0.0;
		fTF_D2 = 0.0375 * fSUM1 + (-0.0225) * fIn - (-1) * fOut;
		fIn = fSUM1;
		fOut = fTF_D2;
	}
	PS_SetPwm1Rate(fTF_D2);

	PS_ExitPwm1General();
}
```

There is also the alternative of representing the digital controller in difference equations, writing them in a PSIM C language block• . Figure 78 shows the simulated circuit with the simple C language block.

Figure 78 - Closed-loop Buck converter with digital controller using the C language block.

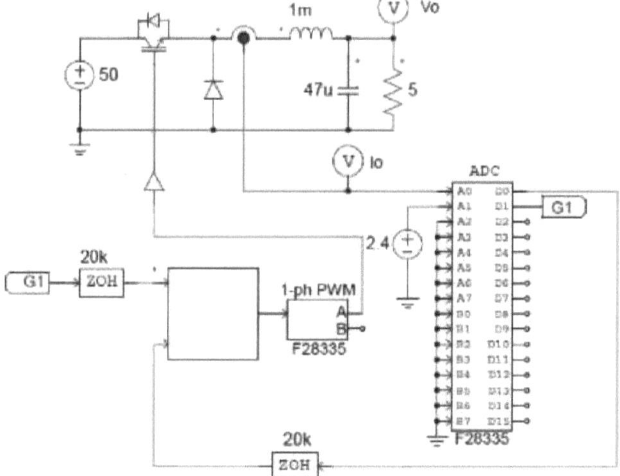

Source: Author's own production.

The code can be seen in Figure 79. Note that the simulation did not use a subtractor to perform the negative feed. This operation was carried out by the code and is represented by the variable e0. The parameters ki and Ti are the gain and time constant of the PI controller, respectively. Ta is the sampling period, being the inverse of the sampling frequency.

Figure 79 - Simple C language block with code written for digital control.

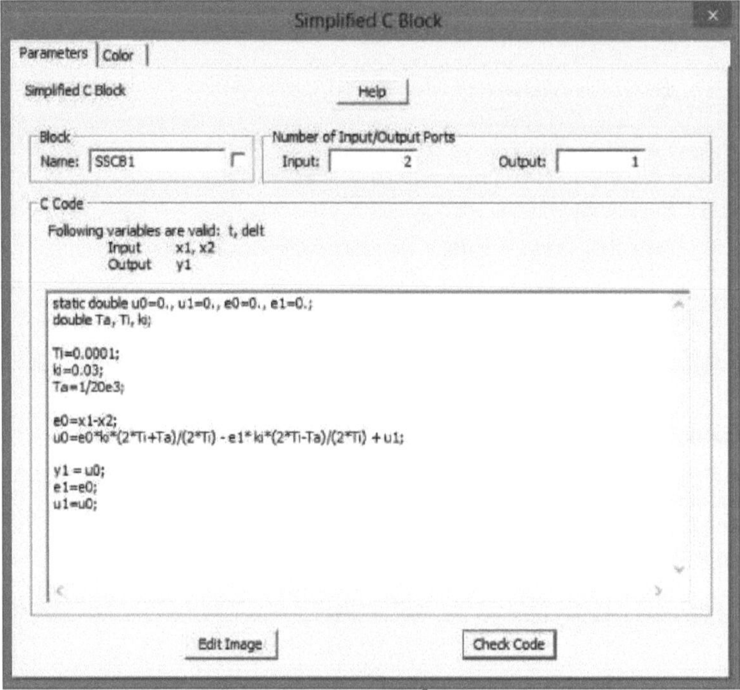

Source: PSIM .®

Figure 80 shows the output waveforms of the simulated converter with the simple C language block. It can be seen that the result was the same as the other methods.

Figure 80 - Buck converter output waveforms with digital controller using the simple C language block.

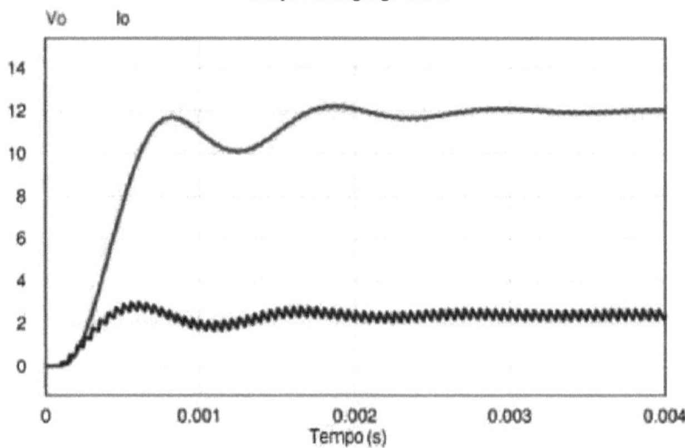

Source: Author's own production.

The code generated by SimCoder maintains the code written for the controller, as shown below.

```
interrupt void Task()
{
	DefaultType fTI_ADC1_1, fZOH2, fTI_ADC1, fZOH1;
	PS_EnableIntr();

	fTI_ADC1_1 = PS_GetDcAdc(1);
	fTI_ADC1 = PS_GetDcAdc(0);
	fZOH2 = fTI_ADC1_1;
	fZOH1 = fTI_ADC1;
	{
		static double u0=0., u1=0., e0=0., e1=0.;
		double Ta, Ti, ki;

		Ti=0.0001;
		ki=0.03;
		Ta=1/20e3;

		e0=fZOH2-fZOH1;
		u0=e0*ki*(2*Ti+Ta)/(2*Ti) - e1* ki*(2*Ti-Ta)/(2*Ti) + u1;

		fGblSSCB1 = u0;
		e1=e0;
		u1=u0;
	}
	PS_ExitTimer1Intr();

}
```

Figure 81 shows the output waveforms of the Buck converter with the controllers of the three methods used to generate code. It can be seen that the waveforms are the same.

Figure 81 - Buck converter output waveforms in closed loop for the three digital control methods.

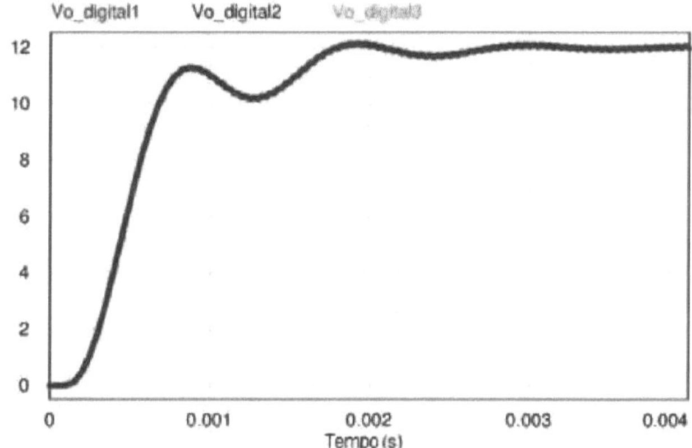

Source: Author's own production.

5.2. FULL BRIDGE INVERTER

An A/D converter was used to simulate the full-bridge inverter in order to obtain voltage and current readings for the load. For this purpose, voltage and current sensors were used, with gains of 1/256 and 1/64 respectively. Channels A0 were used for voltage and A1 for current, both with unity gains. An offset level of 1.5V was added to both channels to allow the A/D converter to be in DC mode. The simulated inverter can be seen in Figure 82.

Figure 82 - Simulation circuit and DSC settings for the full bridge inverter.

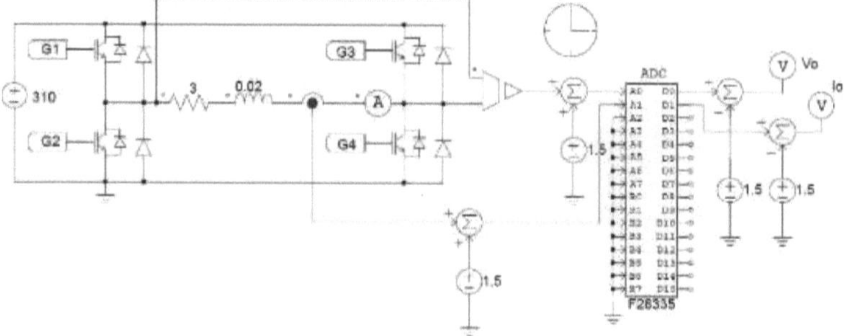

Source: Author's own production.

It can be seen that no filter was used between the sensor outputs and the A/D converter. Filters could have been used to output a sinusoidal waveform for the voltage, to make the simulation more realistic.

Figure 83 shows the configured parameters of the PWM block.

Figure 83 - Settings window for the PWM block used in the full bridge inverter simulation.

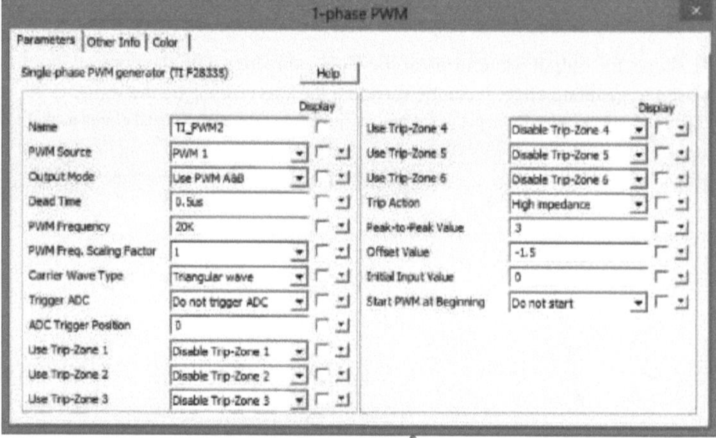

Source: PSIM.®

Two single-phase PWM blocks were used to control the switches of the Full Bridge Inverter, using the PWM1 and PWM2 channels. Both were configured for 20kHz frequency, 0).5LIS dead time, 3V peak-to-peak value, - 1.5V offset value and not to start up. As input for the PWM modulators, two sawtooth sources were used, with a frequency of 60Hz and a maximum value of 360V, and a sine function, obtaining a sine wave ranging from -1 to 1V. For the second block, the sawtooth signal was configured with an offset of 180, allowing the reference sine to be offset by 180° from the sine used for the first PWM modulator.

Start PWM and Stop PWM blocks were used to drive the generators, as was done for the Buck converter, using a digital input block. Figure 84 shows the components mentioned.

Figure 84 - Full bridge inverter control circuit

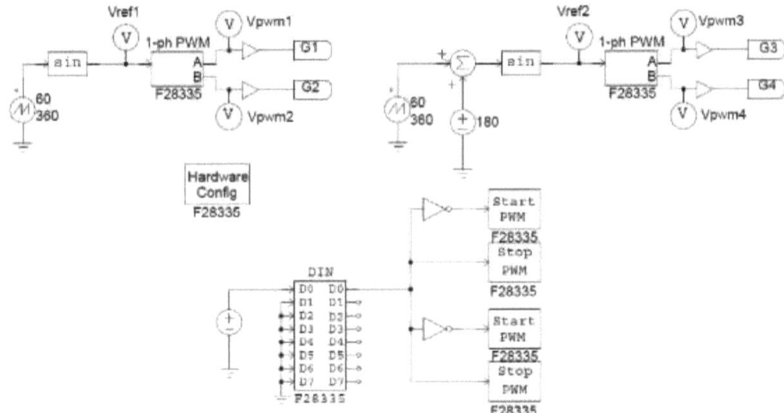

Source: Author's own production.

The waveforms obtained are shown in Figure 85.

Figure 85 - Full bridge inverter output voltage and current

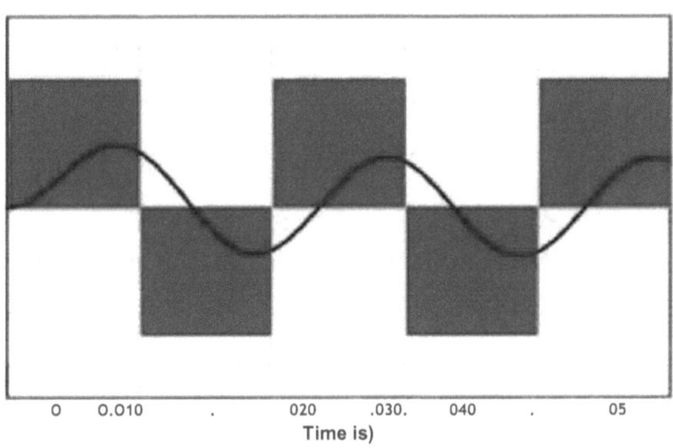

Source: Author's own production.

Based on the simulation carried out, the C language code shown below was generated.

```
/*****************************************************************************
// This code is created by SimCoder Version 9.1 for TI F28335 Hardware Target
//
// SimCoder is copyright by Powersim Inc., 2009-2011
//
// Date: November 02, 2014 17:02:25
******************************************************************************
/
#include      <math.h>
#include      "PS_bios.h"
typedef float DefaultType;
#define       GetCurTime() PS_GetSysTimer()

interrupt void Task();
void Task_1();

DefaultType  fGblV13 = 0.0;
DefaultType  fGblV17 = 0.0;
DefaultType  fGblVo = 0.0;
DefaultType  fGblIo = 0.0;
typedef struct {
       unsigned long tmLow;
       unsigned long tmHigh;
}      _CBigTime;

_CBigTime GetBigTime(void)
```

```
{
	static _CBigTime tm = {0,0};
	unsigned long curTime = GetCurTime();
	if (curTime < tm.tmLow)
		tm.tmHigh++;
	tm.tmLow = curTime;
	return tm;
}

interrupt void Task()
{
	DefaultType fVSAW4, fVDC6, fSUMP1, fSIN3, fVSAW2, fSIN1;
	PS_EnableIntr();

	{
		static unsigned long period = (unsigned long)(150000000L / 50);
		static float fPeriod = ((float)50) / 150000000L;
		static _CBigTime tmCarrierStart = {0, 0};
		_CBigTime tm = GetBigTime();
		unsigned long tmp1, tmp2;
		tmp1 = tm.tmLow - tmCarrierStart.tmLow;
		tmp2 = tm.tmHigh - tmCarrierStart.tmHigh;
		if (tm.tmLow > tmCarrierStart.tmLow)
			tmp2++;
		if (tmp2 || (!tmp2 && (tmp1 >= period))) {
			tmp1 = tmCarrierStart.tmLow + period;
			if ((tmp1 < tmCarrierStart.tmLow) || (tmp1 < period))
				tmCarrierStart.tmHigh++;
			tmCarrierStart.tmLow = tmp1;
		}
		tmp1 = tm.tmLow - tmCarrierStart.tmLow;
		fVSAW4 = 360 * tmp1 * fPeriod;
	}
	fVDC6 = 180;
	fSUMP1 = fVSAW4 + fVDC6;
	fSIN3 = sin(fSUMP1 * (3.14159265 / 180.));
	PS_SetPwm2Rate(fSIN3);
	{
		static unsigned long period = (unsigned long)(150000000L / 50);
		static float fPeriod = ((float)50) / 150000000L;
		static _CBigTime tmCarrierStart = {0, 0};
		_CBigTime tm = GetBigTime();
		unsigned long tmp1, tmp2;
		tmp1 = tm.tmLow - tmCarrierStart.tmLow;
		tmp2 = tm.tmHigh - tmCarrierStart.tmHigh;
		if (tm.tmLow > tmCarrierStart.tmLow)
			tmp2++;
		if (tmp2 || (!tmp2 && (tmp1 >= period))) {
			tmp1 = tmCarrierStart.tmLow + period;
			if ((tmp1 < tmCarrierStart.tmLow) || (tmp1 < period))
				tmCarrierStart.tmHigh++;
			tmCarrierStart.tmLow = tmp1;
		}
		tmp1 = tm.tmLow - tmCarrierStart.tmLow;
		fVSAW2 = 360 * tmp1 * fPeriod;
	}
	fSIN1 = sin(fVSAW2 * (3.14159265 / 180.));
```

```
#ifdef _DEBUG
	fGblV13 = fSIN1;
#endif
	PS_SetPwm1Rate(fSIN1);
#ifdef _DEBUG
	fGblV17 = fSIN3;
#endif
	PS_ExitPwm2General();
}

void Task_1()
{
	DefaultType fTI_DIN1, fNOT5, fNOT9, fTI_ADC1, fVDC9, fSUM1, fTI_ADC1_1, 
fVDC10, fSUM2;

	fTI_DIN1 = (PS_GetDigitInA() & ((Uint32)1 << 5)) ? 1 : 0;
	fTI_ADC1 = PS_GetDcAdc(0);
	fTI_ADC1_1 = PS_GetDcAdc(1);
	fNOT5 = !fTI_DIN1;
	if (fNOT5 > 0)
	{
		PS_StartPwm(1);
	}
	if (fTI_DIN1 != 0)
	{
		PS_StopPwm(2);
	}
	fNOT9 = !fTI_DIN1;
	if (fNOT9 > 0)
	{
		PS_StartPwm(2);
	}
	if (fTI_DIN1 != 0)
	{
		PS_StopPwm(1);
	}
	fVDC9 = 1.5;
	fSUM1 = fTI_ADC1 - fVDC9;
#ifdef _DEBUG
	fGblVo = fSUM1;
#endif
	fVDC10 = 1.5;
	fSUM2 = fTI_ADC1_1 - fVDC10;
#ifdef _DEBUG
	fGblIo = fSUM2;
#endif
}

void Initialize(void)
{
	PS_SysInit(30, 10);
	PS_StartStopPwmClock(0);
	PS_InitTimer(0, 0xffffffff);
	PS_InitPwm(2, 1, 20000*1, (0.5e-6)*1e6, PWM_TWO_OUT, 34115);	// pwnNo, 
waveType, frequency, deadtime, outtype
	PS_SetPwmPeakOffset(2, 3, (-1.5), 1.0/3);
	PS_SetPwmIntrType(2, ePwmNoAdc, 1, 0);
```

```
        PS_SetPwmVector(2, ePwmNoAdc, Task);
        PS_SetPwm2Rate(0);
        PS_StopPwm(2);

        PS_InitPwm(1, 1, 20000*1, (0.5e-6)*1e6, PWM_TWO_OUT, 34115);      // pwnNo,
waveType, frequency, deadtime, outtype
        PS_SetPwmPeakOffset(1, 3, (-1.5), 1.0/3);
        PS_SetPwmIntrType(1, ePwmNoAdc, 1, 0);
        PS_SetPwm1Rate(0);
        PS_StopPwm(1);

        PS_ResetAdcConvSeq();
        PS_SetAdcConvSeq(eAdcCascade, 0, 1);
        PS_SetAdcConvSeq(eAdcCascade, 1, 1);
        PS_AdcInit(0, !0);

        PS_InitDigitIn(5, 100);

        PS_StartStopPwmClock(1);
}

void main()
{
        Initialize();
        PS_EnableIntr();   // Enable Global interrupt INTM
        PS_EnableDbgm();
        for (;;) {
                Task_1();
        }
}
```

5.3. THREE-PHASE INVERTER

To simulate the three-phase inverter, the circuit shown in Figure 86 was used, where the Vin source is 24 V and the three-phase load has R1 = R2 = R3 = 5Ω and L1 = L2 = L3 = 0.01H. To measure the inverter's output, three channels of an A/D converter were used, measuring the phase voltages on the three-phase load. The sensors have gains of 1/16, so that the voltage value after the band-pass filters is between -1.5V and 1.5V of the A/D converter's AC mode. However, offset levels of 1.5V are added to use the A/D converter in DC mode. To compensate for this sensor gain, the A/D converter was configured to have a gain of 16V for the three channels used.

Figure 86 - Three-phase inverter power circuit.

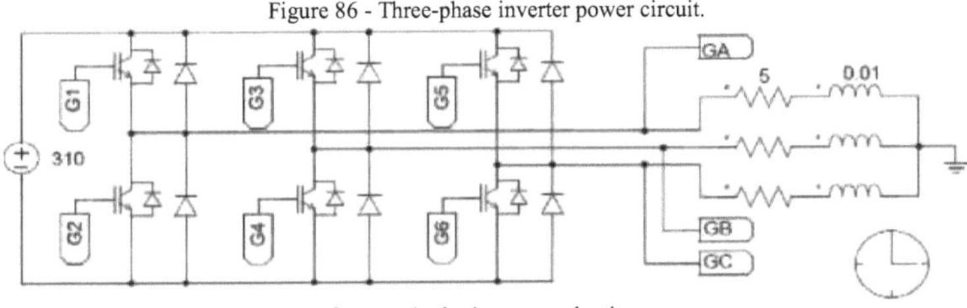

Source: Author's own production.

A three-phase PWM block was used to control this inverter, which was configured for 0.5LIS dead time, 20kHz frequency, 3V peak carrier voltage and -1.5V offset. Three 1V amplitude sinusoidal signals were used as reference signals, offset by 120° from each other. Figure 87 shows the PWM generator and a circuit to activate the generator, similar to the one shown for the Buck converter.

Figure 87 - Three-phase inverter control circuit.

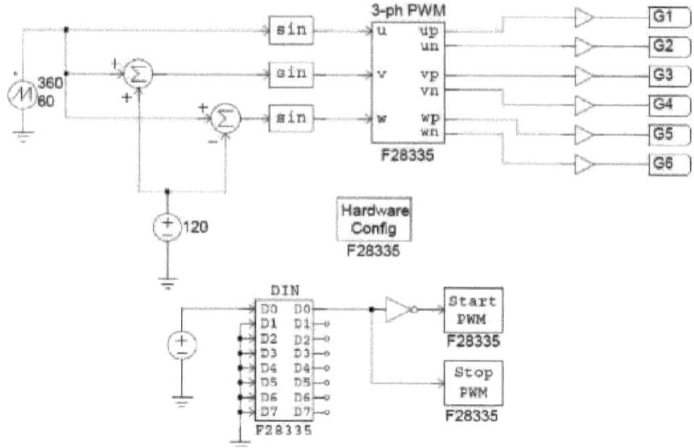

Source: Author's own production.

Figure 88 shows the circuit for analogue readings via the A/D converter.

Figure 88 - A/D converter used for the three-phase inverter.

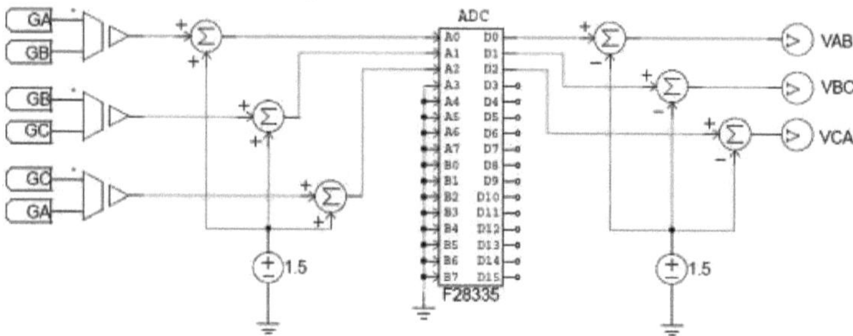

The waveforms of the voltages read by the A/D converter can be seen in Figure 89.

Figure 89 - Output voltage of the three phases of the three-phase inverter.

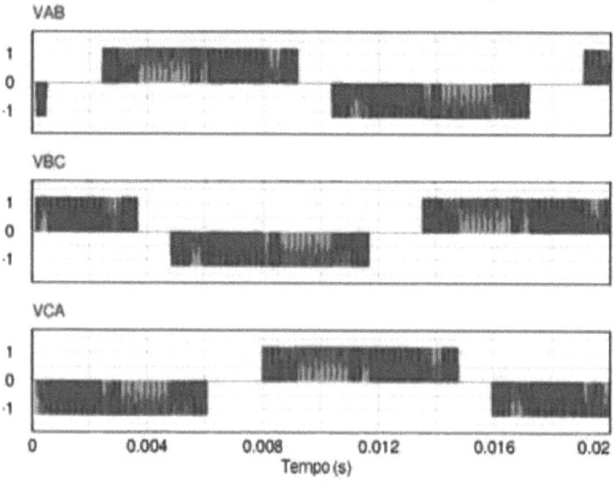

VAB
1
0
-1
VBC
1
0
-1
VCA
1
0
-1
0 0.004 0.008 0.012 0.016 0.02
Tempo (s)

Source: Author's own production.

The simulated circuit was used to generate the C language code shown below.

```
/*******************************************************************************
// This code is created by SimCoder Version 9.1 for TI F28335 Hardware Target
//
// SimCoder is copyright by Powersim Inc., 2009-2011
//
// Date: November 02, 2014 17:42:06
*******************************************************************************
/
#include     <math.h>
#include     "PS_bios.h"
typedef float DefaultType;
#define      GetCurTime() PS_GetSysTimer()

interrupt void Task();
void Task_1();

DefaultType  fGblVAB = 0.0;
DefaultType  fGblVBC = 0.0;
DefaultType  fGblVCA = 0.0;
typedef struct {
	unsigned long tmLow;
	unsigned long tmHigh;
}	_CBigTime;

_CBigTime GetBigTime(void)
{
	static _CBigTime tm = {0,0};
	unsigned long curTime = GetCurTime();
	if (curTime < tm.tmLow)
		tm.tmHigh++;
	tm.tmLow = curTime;
	return tm;
}

interrupt void Task()
{
	DefaultType fVSAW1, fSIN1, fVDC1, fSUMP1, fSIN2, fSUM1, fSIN3;
	PS_EnableIntr();

	{
		static unsigned long period = (unsigned long)(150000000L / 60);
		static float fPeriod = ((float)60) / 150000000L;
		static _CBigTime tmCarrierStart = {0, 0};
		_CBigTime tm = GetBigTime();
		unsigned long tmp1, tmp2;
		tmp1 = tm.tmLow - tmCarrierStart.tmLow;
		tmp2 = tm.tmHigh - tmCarrierStart.tmHigh;
		if (tm.tmLow > tmCarrierStart.tmLow)
			tmp2++;
		if (tmp2 || (!tmp2 && (tmp1 >= period))) {
			tmp1 = tmCarrierStart.tmLow + period;
			if ((tmp1 < tmCarrierStart.tmLow) || (tmp1 < period))
```

```
                        tmCarrierStart.tmHigh++;
                    tmCarrierStart.tmLow = tmp1;
                }
                tmp1 = tm.tmLow - tmCarrierStart.tmLow;
                fVSAW1 = 360 * tmp1 * fPeriod;
        }
        fSIN1 = sin(fVSAW1 * (3.14159265 / 180.));
        fVDC1 = 120;
        fSUMP1 = fVSAW1 + fVDC1;
        fSIN2 = sin(fSUMP1 * (3.14159265 / 180.));
        fSUM1 = fVSAW1 - fVDC1;
        fSIN3 = sin(fSUM1 * (3.14159265 / 180.));
        PS_SetPwm3ph1Uvw(fSIN1, fSIN2, fSIN3);
        PS_ExitPwm1General();
}

void Task_1()
{
        DefaultType fTI_ADC1, fVDC6, fSUM2, fTI_ADC1_1, fVDC8, fSUM3, fTI_ADC1_2,
fVDC9, fSUM4;
        DefaultType fTI_DIN1, fNOT1;

        fTI_ADC1 = PS_GetDcAdc(0);
        fTI_ADC1_1 = PS_GetDcAdc(1);
        fTI_ADC1_2 = PS_GetDcAdc(2);
        fTI_DIN1 = (PS_GetDigitInA() & ((Uint32)1 << 10)) ? 1 : 0;
        fVDC6 = 1.5;
        fSUM2 = fTI_ADC1 - fVDC6;
#ifdef _DEBUG
        fGblVAB = fSUM2;
#endif
        fVDC8 = 1.5;
        fSUM3 = fTI_ADC1_1 - fVDC8;
#ifdef _DEBUG
        fGblVBC = fSUM3;
#endif
        fVDC9 = 1.5;
        fSUM4 = fTI_ADC1_2 - fVDC9;
#ifdef _DEBUG
        fGblVCA = fSUM4;
#endif
        fNOT1 = !fTI_DIN1;
        if (fNOT1 > 0)
        {
                PS_StartPwm3ph1();
        }
        if (fTI_DIN1 != 0)
        {
                PS_StopPwm3ph1();
        }
}

void Initialize(void)
{
        PS_SysInit(30, 10);
        PS_StartStopPwmClock(0);
        PS_InitTimer(0, 0xffffffff);
```

```
        PS_InitPwm3ph(1, 1, 20000*1, (0.5e-6)*1.e6, 1811); // pwnNo, waveType,
frequency, deadtime
        PS_SetPwm3phPeakOffset(1, 3, (-1.5), 1.0/3);
        PS_SetPwm3ph1AdcIntr(ePwmNoAdc, 1, 0);
        PS_SetPwm3ph1Vector(ePwmNoAdc, Task);
        PS_SetPwm3ph1Uvw(0, 0, 0);
        PS_StopPwm3ph1();

        PS_ResetAdcConvSeq();
        PS_SetAdcConvSeq(eAdcCascade, 0, 1);
        PS_SetAdcConvSeq(eAdcCascade, 1, 1);
        PS_SetAdcConvSeq(eAdcCascade, 2, 1);
        PS_AdcInit(0, !0);

        PS_InitDigitIn(10, 100);

        PS_StartStopPwmClock(1);
}

void main()
{
        Initialize();
        PS_EnableIntr();    // Enable Global interrupt INTM
        PS_EnableDbgm();
        for (;;) {
            Task_1();
        }
}
```

5.4. NPC INVERTER

Figure 90 shows the simulated NPC inverter. This inverter has three arms with four switches on each. Each switch requires a PWM channel for its control, making twelve channels necessary. The load used in the simulation has the same values as those used for the three-phase inverter.

Two three-phase PWM blocks were used to control this inverter, one for the odd-numbered switches and one for the even-numbered switches, as shown in the figure above. The PWM modulators were configured for frequencies of 20kHz, with a triangular carrier and a dead time of 0.5us. Only the peak values of the carriers were set differently, with the first modulator set to 1V peak and no offset, and the second to 1V peak and -1 offset.

Figure 90 - Simulation circuit and DSC settings for the simulated NPC inverter.

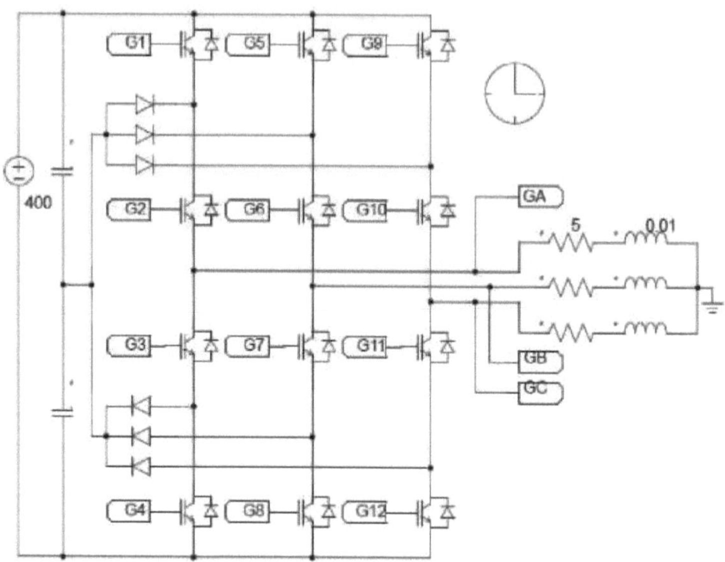

Source: Author's own production.

This modulation strategy is called IPD (In-Phase Disposition) because the two carriers are in phase. Another similar type of modulation is POD (Phase Opposition Disposition), which also has two carriers, one positive and one negative. In this type of modulation the carriers are out of phase by 180° and it will not be used as there is a risk of triggering two switches on the same arm simultaneously when the carriers cross at zero.

Three reference sine waves were used, offset by 120° from each other. Each signal was connected to the two three-phase PWM blocks, one with a positive carrier and one with a negative carrier, as shown in Figure 91.

Figure 91 - Simulated NPC inverter control.

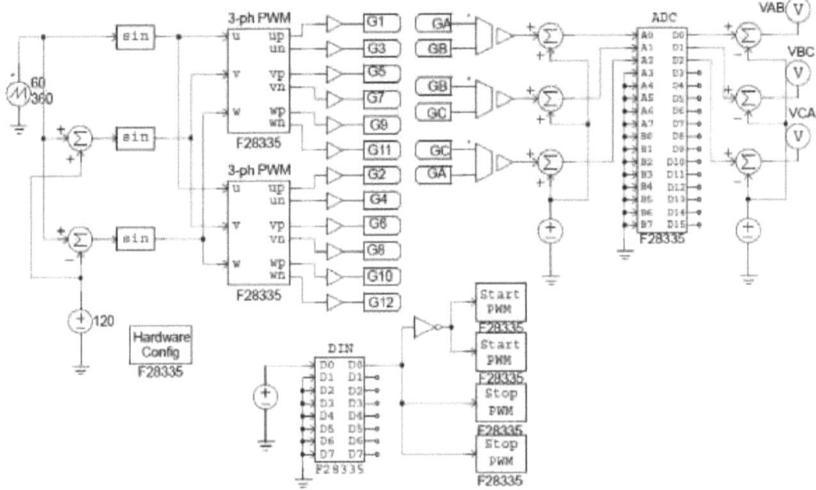

Source: Author's own production.

The rest of the components used are the same as those used in the previous simulations. The gain of the sensors is 1/400, for reading the A/D converter in DC mode. For the circuit to activate the PWM generators, two Start PWM blocks and two Stop PWM blocks were required.

The output of the NPC is a signal with three levels, as shown in Figure 92. The figure clearly identifies the levels corresponding to Vin, Vin/2, -Vin and -Vin/2. In the simulation, Vin corresponds to 400V, but due to the gain of the sensors, the levels were changed to 1V, 0.5V, -1V and -0.5V.

Figure 92 - NPC output voltage waveforms.

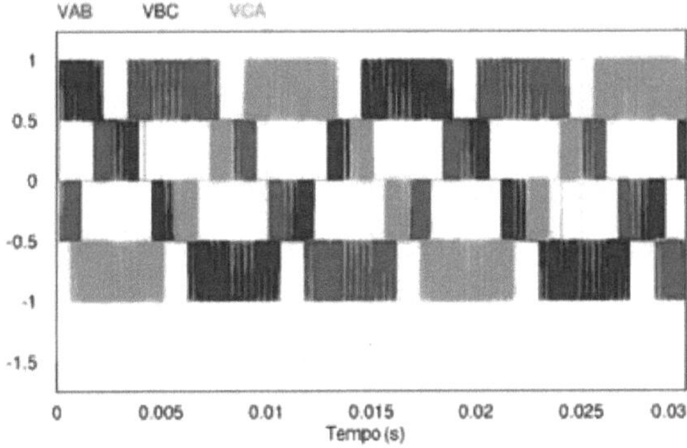

Source: Author's own production.

The code below was generated from the NPC simulation.

```
/*******************************************************************************
// This code is created by SimCoder Version 9.1 for TI F28335 Hardware Target
//
// SimCoder is copyright by Powersim Inc., 2009-2011
//
// Date: November 02, 2014 18:25:26
*******************************************************************************
/
#include     <math.h>
#include     "PS_bios.h"
typedef float DefaultType;
#define      GetCurTime() PS_GetSysTimer()

interrupt void Task();
void Task_1();

DefaultType  fGblVAB = 0.0;
DefaultType  fGblVBC = 0.0;
DefaultType  fGblVCA = 0.0;
typedef struct {
      unsigned long tmLow;
      unsigned long tmHigh;
}     _CBigTime;

_CBigTime GetBigTime(void)
{
      static _CBigTime tm = {0,0};
```

```
	unsigned long curTime = GetCurTime();
	if (curTime < tm.tmLow)
		tm.tmHigh++;
	tm.tmLow = curTime;
	return tm;
}

interrupt void Task()
{
	DefaultType fVSAW1, fSIN1, fVDC2, fSUMP1, fSIN2, fSUM1, fSIN3;
	PS_EnableIntr();

	{
		static unsigned long period = (unsigned long)(150000000L / 60);
		static float fPeriod = ((float)60) / 150000000L;
		static _CBigTime tmCarrierStart = {0, 0};
		_CBigTime tm = GetBigTime();
		unsigned long tmp1, tmp2;
		tmp1 = tm.tmLow - tmCarrierStart.tmLow;
		tmp2 = tm.tmHigh - tmCarrierStart.tmHigh;
		if (tm.tmLow > tmCarrierStart.tmLow)
			tmp2++;
		if (tmp2 || (!tmp2 && (tmp1 >= period))) {
			tmp1 = tmCarrierStart.tmLow + period;
			if ((tmp1 < tmCarrierStart.tmLow) || (tmp1 < period))
				tmCarrierStart.tmHigh++;
			tmCarrierStart.tmLow = tmp1;
		}
		tmp1 = tm.tmLow - tmCarrierStart.tmLow;
		fVSAW1 = 360 * tmp1 * fPeriod;
	}
	fSIN1 = sin(fVSAW1 * (3.14159265 / 180.));
	fVDC2 = 120;
	fSUMP1 = fVSAW1 + fVDC2;
	fSIN2 = sin(fSUMP1 * (3.14159265 / 180.));
	fSUM1 = fVSAW1 - fVDC2;
	fSIN3 = sin(fSUM1 * (3.14159265 / 180.));
	PS_SetPwm3ph1Uvw(fSIN1, fSIN2, fSIN3);
	PS_SetPwm3ph2Uvw(fSIN1, fSIN2, fSIN3);
	PS_ExitPwm1General();
}

void Task_1()
{
	DefaultType fTI_DIN1, fNOT1, fTI_ADC1, fVDC5, fSUM2, fTI_ADC1_1, fSUM3,
fTI_ADC1_2, fSUM4;

	fTI_DIN1 = (PS_GetDigitInA() & ((Uint32)1 << 20)) ? 1 : 0;
	fTI_ADC1 = PS_GetDcAdc(0);
	fTI_ADC1_1 = PS_GetDcAdc(1);
	fTI_ADC1_2 = PS_GetDcAdc(2);
	fNOT1 = !fTI_DIN1;
	if (fNOT1 > 0)
	{
		PS_StartPwm3ph1();
	}
	if (fNOT1 > 0)
```

```
        {
                PS_StartPwm3ph2();
        }
        if (fTI_DIN1 != 0)
        {
                PS_StopPwm3ph1();
        }
        if (fTI_DIN1 != 0)
        {
                PS_StopPwm3ph2();
        }
        fVDC5 = 1.5;
        fSUM2 = fTI_ADC1 - fVDC5;
#ifdef _DEBUG
        fGblVAB = fSUM2;
#endif
        fSUM3 = fTI_ADC1_1 - fVDC5;
#ifdef _DEBUG
        fGblVBC = fSUM3;
#endif
        fSUM4 = fTI_ADC1_2 - fVDC5;
#ifdef _DEBUG
        fGblVCA = fSUM4;
#endif
}

void Initialize(void)
{
        PS_SysInit(30, 10);
        PS_StartStopPwmClock(0);
        PS_InitTimer(0, 0xffffffff);
        PS_InitPwm3ph(1, 1, 20000*1, (0.5e-6)*1.e6, 36118); // pwnNo, waveType,
frequency, deadtime
        PS_SetPwm3phPeakOffset(1, 1, 0, 1.0/1);
        PS_SetPwm3ph1AdcIntr(ePwmNoAdc, 1, 0);
        PS_SetPwm3ph1Vector(ePwmNoAdc, Task);
        PS_SetPwm3ph1Uvw(0, 0, 0);
        PS_StartPwm3ph1();

        PS_InitPwm3ph(2, 1, 20000*1, (0.5e-6)*1.e6, 36118); // pwnNo, waveType,
frequency, deadtime
        PS_SetPwm3phPeakOffset(2, 1, (-1), 1.0/1);
        PS_SetPwm3ph2AdcIntr(ePwmNoAdc, 1, 0);
        PS_SetPwm3ph2Uvw(0, 0, 0);
        PS_StartPwm3ph2();

        PS_ResetAdcConvSeq();
        PS_SetAdcConvSeq(eAdcCascade, 0, 1);
        PS_SetAdcConvSeq(eAdcCascade, 1, 1);
        PS_SetAdcConvSeq(eAdcCascade, 2, 1);
        PS_AdcInit(0, !0);

        PS_InitDigitIn(20, 100);

        PS_StartStopPwmClock(1);
}
```

```
void main()
{
        Initialize();
        PS_EnableIntr();   // Enable Global interrupt INTM
        PS_EnableDbgm();
        for (;;) {
                Task_1();
        }
}
```

CHAPTER 6

6. EXPERIMENTAL RESULTS

To carry out tests using the TMS320F28335, the Code Composer Studio software, version 5.1, will be used to compile and debug the generated codes, observing the functionalities of the SimCoder elements.

6.1 TESTS WITH DSC

Firstly, open a project, following the steps explained in Chapter 4. To debug the code for the DSC, you must create a target configuration file, in which you must choose which device will be used. To create this file, go to File>New>Target Configuration File. When you open this file, select the type of connection. In this case, the Texas Instruments XDS100v1 USB Emulator will be used. Then select the device used, in this case the DSC TMS320F28335, and save it. Figure 93 shows the aforementioned settings.

Figure 93 - Device configuration by CCSv5.1

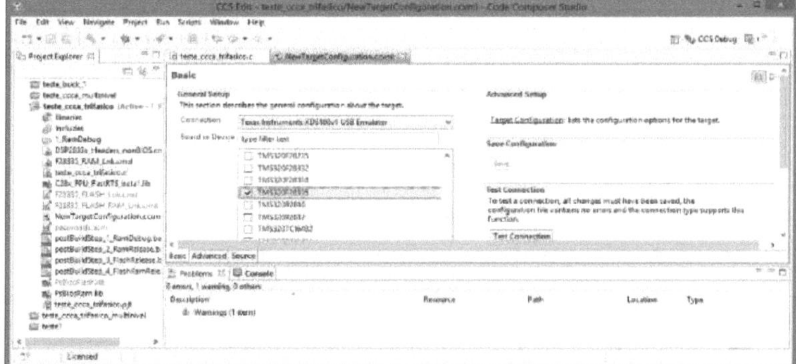

Figure 93 shows the Test Connection icon, which allows the connection to be tested. If an error occurs, the CCS must be closed, the USB port used changed and opened again.

To begin the tests, the simulated circuit and code shown in Chapter 4, in Figure 61, were used. The circuit uses a value obtained by the A/D converter to determine the cyclic ratio of the single-phase PWM modulator. The PWM block is driven by the Start PWM and Stop PWM components, which in turn are driven by a digital input. From the generated C language code, the code can then be debugged for the DSC.

A potentiometer was used to vary the voltage at input A0, corresponding to the A/D converter input configured in the simulation, allowing a variation of approximately 0 to 3.5V. Figure 94 shows the PWM signal generated by the DSC, measured at the GPIO00 port, and the A/D converter input signal, which corresponds to the PWM cyclic ratio, measured at the DSC's ADCA0 port.

Figure 94 - PWM signal and A/D converter input voltage readings.

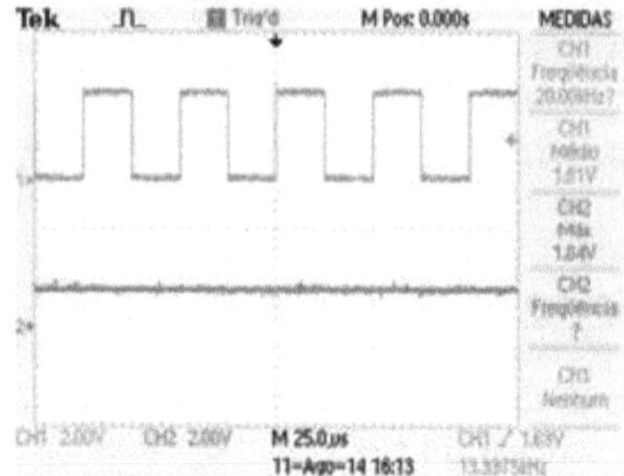

Source: Author's own production.

The single-phase PWM modulator generates two complementary channels, as seen in Figure 95. The second PWM channel can be measured on GPIO port 01.

Figure 95 - Complementary PWM signals obtained.

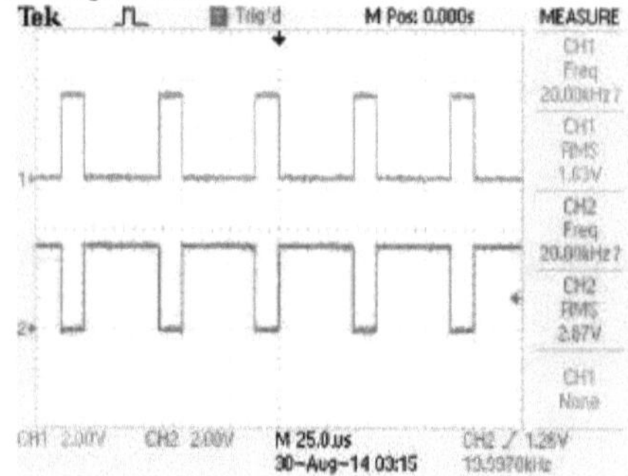

Source: Author's own production.

In Figure 96, you can see from the dotted lines that the dead time between the two PWM signals generated is actually 0.5gs, checking with the value configured in the component through the PSIM simulation.

Figure 96 - Dead time between PWM channels.

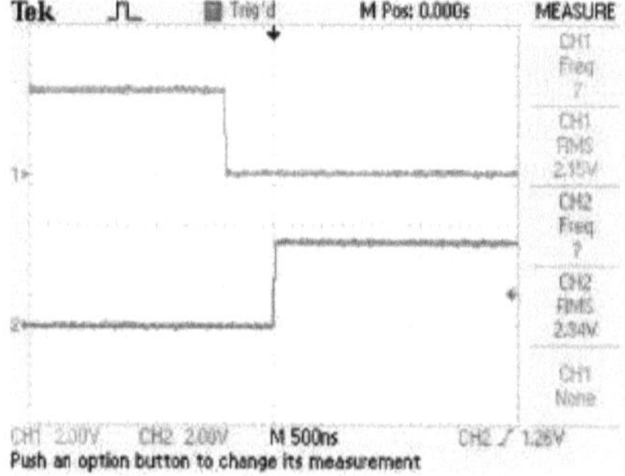

Source: Author's own production.

Using a sinusoidal reference for the PWM modulator, the waveform shown in Figure 97 was obtained. The reference signal used was inserted into the PSIM simulation itself, which was used to generate the code in C language. The figure shows the PWM signal obtained and the filtered signal. This signal has a frequency of 60Hz and was obtained by applying the PWM signal to a simple RC filter. For better visualisation, a frequency of 1kHz was used for the PWM signal.

Figure 97 - Signals obtained from a sinusoidal reference.

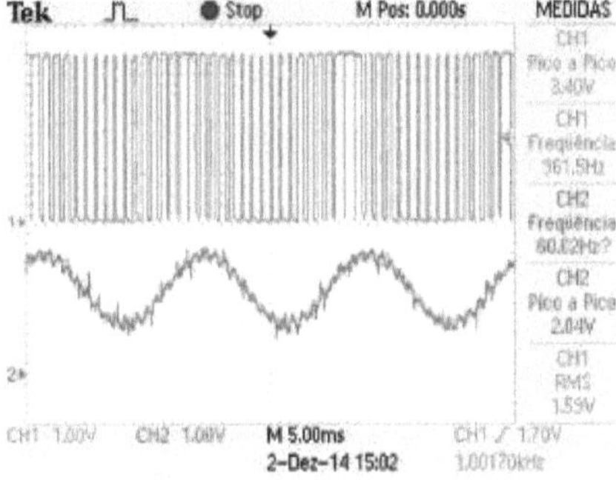

Source: Author's own production.

You can also use the A/D converter to acquire an external reference signal for the PWM generator. Figure 98 shows a reference sine wave supplied by a signal generator source and the PWM signal generated. It can be seen that the cyclic ratio of the PWM signal varies according to the reference signal.

Figure 98 - External reference signal and generated PWM signal.

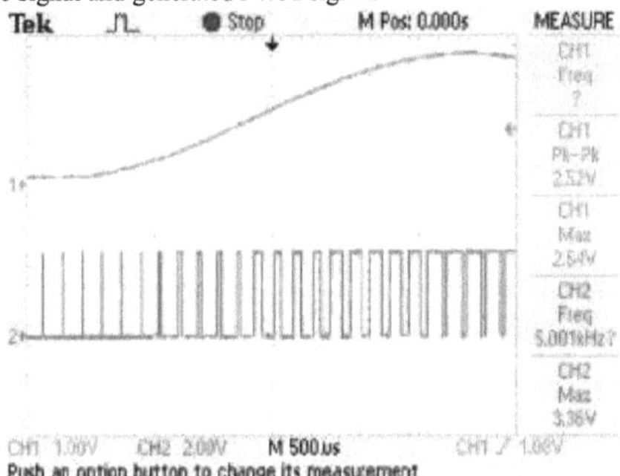

Source: Author's own production.

Figure 99 shows two signals generated by a PWM signal with a sinusoidal reference after an RC filter. The signals were obtained from two channels of the same PWM generator, so the signals are complementary, generating sinusoids offset by 180°.

Figure 99 - Signals generated by complementary PWM signals, after filtering.

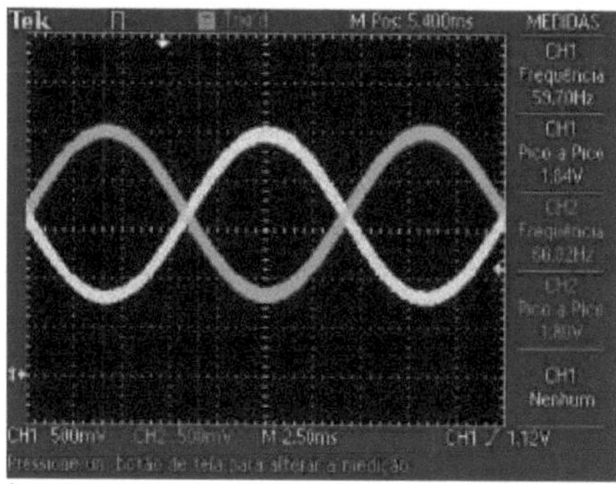

Source: Author's own production.

Figure 100 shows three signals obtained from a PWM generator with a sinusoidal reference, offset by 120°, in order to check the operation of the three-phase PWM generator.

Figure 100 - Signals obtained from the three-phase PWM modulator, after filtering.

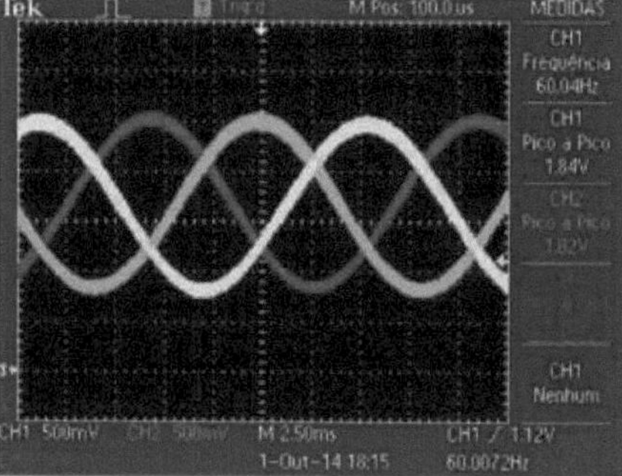

Source: Author's own production.

The three signals were obtained from identical RC filters, using a 100nF capacitor and a 10kΩ resistor.

The lag between the signals could be checked using the oscilloscope's time sliders. Considering that the signals are offset by 120°, the time difference between the peaks of the two phases should be equal to one third of the wave's period. For the 60Hz frequency used, this time difference should be approximately equal to 5.55ms. Figure 101 shows that this time is verified.

Figure 101 - Checking the lag between two phases.

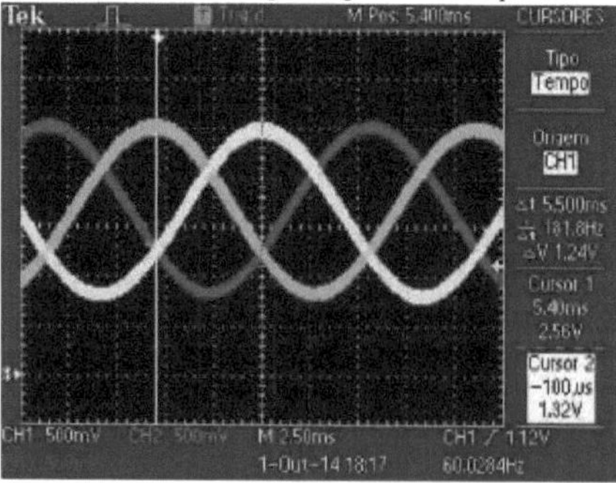

Source: Author's own production.

Finally, a digital input was tested. In the PSIM simulation®, a digital input block was used to trigger the Start and Stop PWM components, determining the operation of the PWM modulator. Figure 102 shows a square waveform obtained by a signal generator and the PWM signal.

Figure 102 - Testing the digital input.

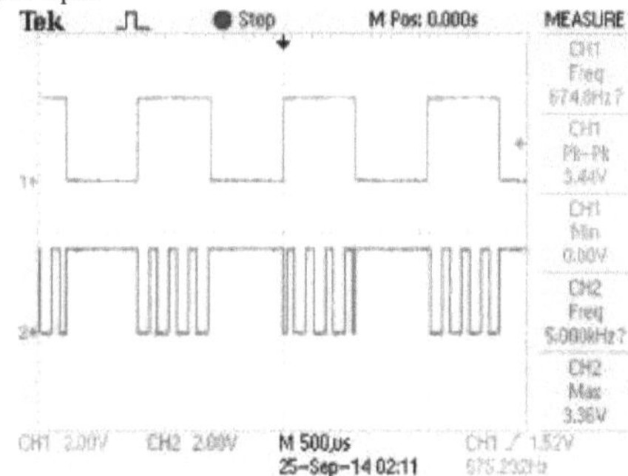

Source: Author's own production.

6.2 SIMULATION OF THE NPCm CONVERTER

The NPCm (Modified Neutral Point Clamped) is an alternative topology to the NPC. Both converters have three arms, with four switches on each arm, but arranged in different ways. Figure 103 shows a representation of an NPCm phase.

Figure 103 - NPCm single-stage topology.

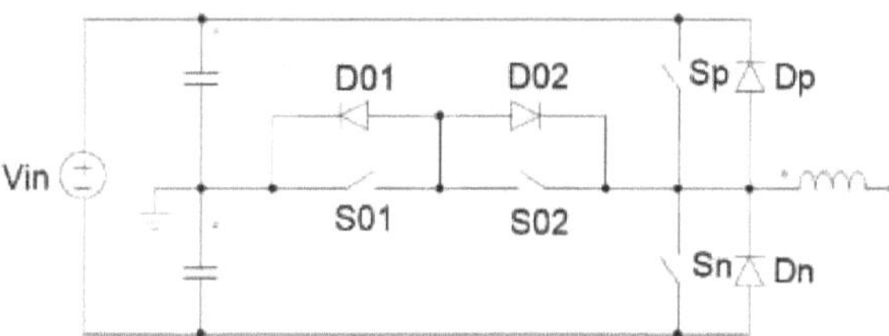

Source: Author's own production.

The main difference between the semiconductors in the two structures is that in the NPCm rectifier the semiconductors S_p, S_n, D_p and D_n are subject to the full DC bus voltage, while all the other devices in both structures block half of this voltage (HEERDT, 2013).

Despite the differences in topology, the modulation strategy of this converter is the same as that used for the NPC, and it has the same characteristics from the point of view of its terminals. However, the NPCm has advantages in terms of the number of components, as it doesn't need two clamping diodes.

After studying this converter, a simulation was carried out using the SimCoder blocks, generating the code in C language for the NPCm converter shown in Figure 104. The converter was already made with the DSC F28335 socket, and the code was implemented using a serial communication cable.

Figure 104 - NPCm used to validate the results.

Source: (HEERDT, 2013)

Figure 105 shows the simulation diagram of the power circuit of the three-phase NPCm converter, made up of the input capacitive filters, switches, output LC filters and 3 Ohm resistive loads. Resonance damping resistors are added in series with some capacitors.

Figure 105 - NPCm converter power circuit.

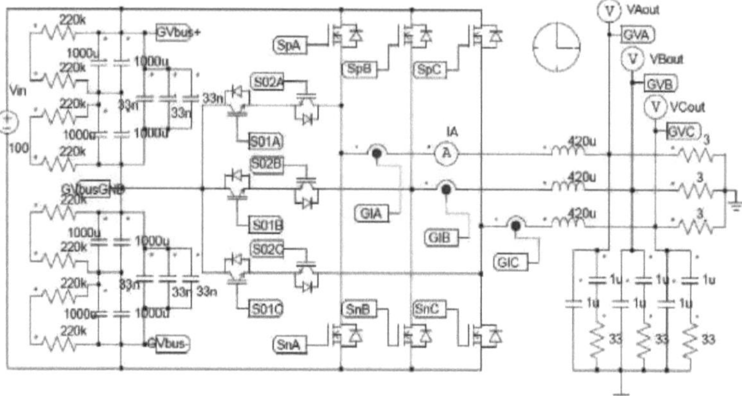

Source: Author's own production.

Six PWM channels are needed to control the converter. As a reference signal, 1V amplitude sinusoids were used, with a frequency of 60Hz, offset by 120° from each other. A Phase Locked Loop (PLL) could be used to generate these reference signals, based on the reading of the converter's output, which is connected to the mains. The PLL is used to synchronise the grid voltages with the voltage produced by the inverter. The PWM blocks are shown in Figure 106.

The modulation used is IDP, where two triangular carriers are used, one positive and one negative. For the converter used, the modulation was designed to perform a positive comparison for the positive carrier and a negative comparison for the negative carrier. Positive comparison is understood to mean that the reference signal is connected to the comparator's positive input and the carrier to the negative input. When the reference signal is connected to the negative input, negative comparison occurs.

Figure 106 - PWM generators for controlling the NPCm.

Source: Author's own production.

However, the SimCoder PWM blocks cannot be configured to perform the negative comparison. The PWM1, PWM5 and PWM6 channels should be configured for the positive carrier and the PWM2, PWM3 and PWM4 channels for the negative carrier. The solution found was to use only one carrier for the six PWM blocks and to offset the reference sinusoid of the channels that should have a negative carrier by 180°. The PWM1 channel settings can be seen in Figure 107.

A small offset level was set to simulate the dead time found in IDP modulation. This ensures that the Sp and Sn switches are not activated simultaneously. This offset level could be disregarded as it distorts the output signals.

Figure 107 - Configurations of one of the PWM generators used to control the NPCm.

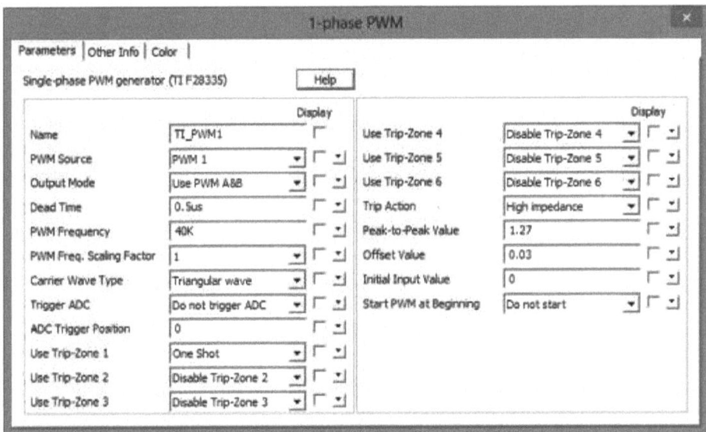

1-phase PWM

Parameters | Other Info | Color

Single-phase PWM generator (TI F28335)   Help

Parameter	Value	Parameter	Value
Name	TI_PWM1	Use Trip-Zone 4	Disable Trip-Zone 4
PWM Source	PWM 1	Use Trip-Zone 5	Disable Trip-Zone 5
Output Mode	Use PWM A&B	Use Trip-Zone 6	Disable Trip-Zone 6
Dead Time	0.5us	Trip Action	High impedance
PWM Frequency	40K	Peak-to-Peak Value	1.27
PWM Freq. Scaling Factor	1	Offset Value	0.03
Carrier Wave Type	Triangular wave	Initial Input Value	0
Trigger ADC	Do not trigger ADC	Start PWM at Beginning	Do not start
ADC Trigger Position	0		
Use Trip-Zone 1	One Shot		
Use Trip-Zone 2	Disable Trip-Zone 2		
Use Trip-Zone 3	Disable Trip-Zone 3		

Source: PSIM.®

Figure 108 shows the reference sine waveforms for phase A of the NPCm converter and the outputs of the PWM generators for the Sp and Sn switches.

Figure 108 - Reference signal and PWM signals generated.

Source: Author's own production.

The NPCm used had several sensors, which were used to take voltage readings of the input values and voltage and current readings of the output values. Eight channels of the A/D converter were used for this purpose. These channels were predetermined, so it was not possible to change them. All the channels were configured for DC mode and to have unity gain. Figure 109 shows the circuit containing the A/D converter and the read values.

It can be seen that for the NPCm's output voltage and current readings, an offset level of 1.5V was inserted and later removed, so that they showed values between 0 and 3V, as required by the A/D converter in DC mode. Note that it is not possible to choose the sampling frequency of the A/D converter in its settings, but it is possible to choose it in the PWM blocks.

Figure 109 - Readings taken by the A/D converter for the NPCm.

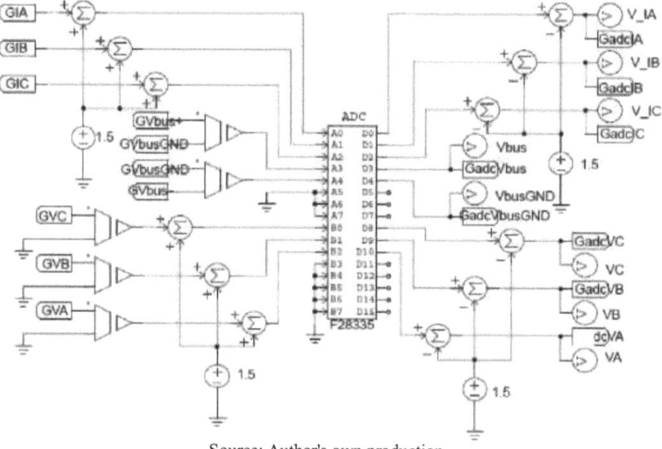

Source: Author's own production.

The NPCm converter used has two buttons used as digital inputs and two LEDs on the DSC itself used as digital outputs. Both GPIO ports were previously defined and cannot be changed. For the digital inputs, ports GPIO24 and GPIO25 were used, and for the outputs, GPIO31 and GPIO34.

The GPIO24 port is used to activate the PWM generators using the Start PWM components, one for each generator. The GPIO25 port is used to stop the PWM modulators, using the Stop PWM components, which will be presented later. Figure 110 shows the digital input and output blocks.

Figure 110 - Digital inputs and outputs used.

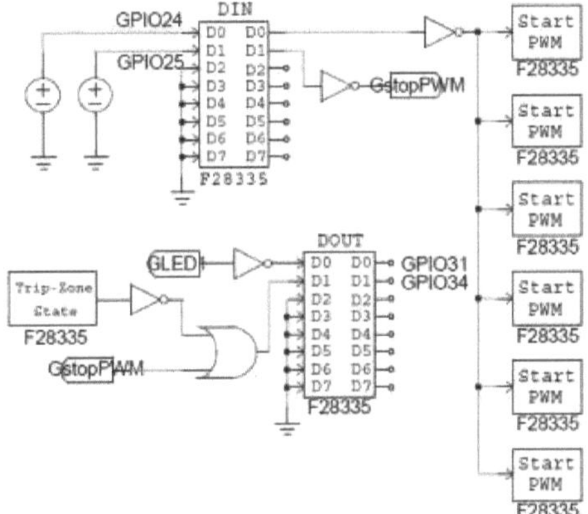

Source: Author's own production.

The LED connected to port GPIO31 is used to indicate that the NPCm protections have been activated. The LED connected to GPIO port 34 indicates whether the button that interrupts the PWM blocks has been activated or whether the Trip-Zone has been activated, via the Trip-Zone State component.

The Trip-Zone was configured for the Trip-Zone 1 channel on the GPIO12 port, which in turn receives a signal from the current sensors to protect against overcurrents. In the PWM modulator configurations shown in Figure 86, Trip-Zone 1 was configured to be used in One-Shot mode, whereby with a pulse on the Trip-Zone input, the PWM generators are switched off and only reactivated if the Start PWM components are activated. The Trip-Zone State block must also be configured for the Trip-Zone 1 channel.

Figure 111 shows the NPCm's protection elements: the TripZone and the comparators that trigger the PWM Stop blocks. The values of Ipk_max, Vbus_max and Vpk_max were determined for the proper protection of the NPCm, and in the simulation these values were divided by the sensor gains.

Figure 111 - NPCm Protection Elements.

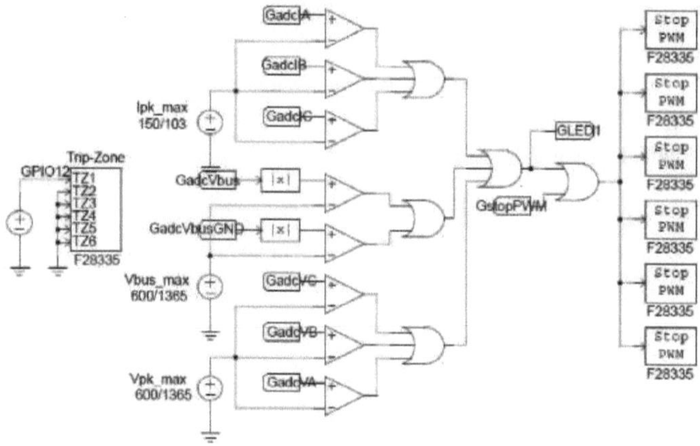

Source: Author's own production.

Figure 112 shows the output voltage and current in phase A of the converter. It can be seen that at each zero crossing there is a small disturbance, due to the dead time between the Sp and Sn switch commands, and between S01 and S02. The disturbances at the peaks of the sinusoids have the same cause.

Figure 112 - Output voltage and current of phase A of the NPCm converter.

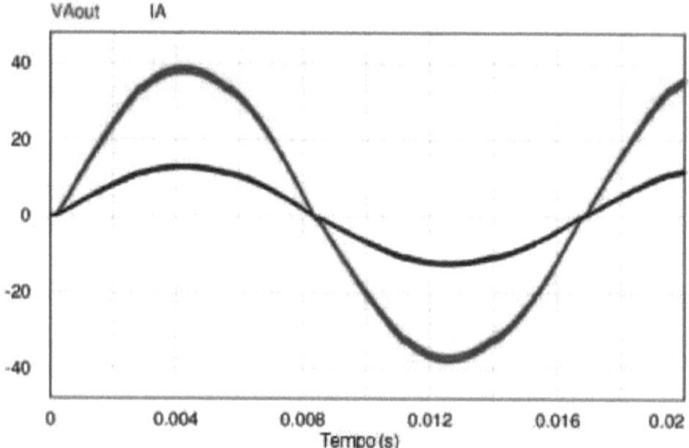

Source: Author's own production.

By removing the offset level from the carriers of the PWM modulators, the disturbances are eliminated, as shown in Figure 113.

Figure 113 - Output voltage and current of the NPCm converter without disturbances.

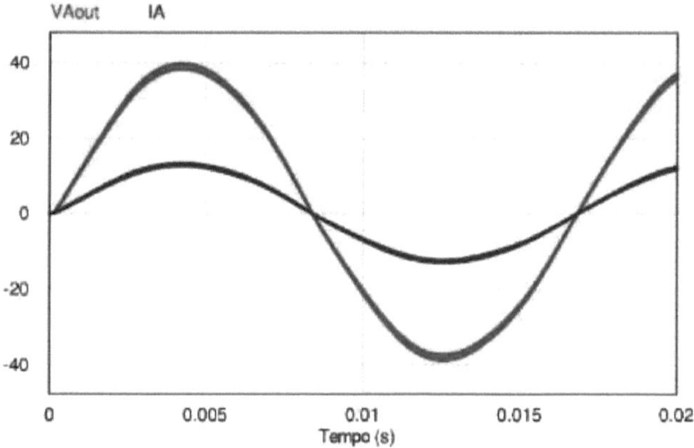

Source: Author's own production.

After simulating the NPCm converter, the code below was generated.

```
/*******************************************************************************
// This code is created by SimCoder Version 9.1 for TI F28335 Hardware Target
//
// SimCoder is copyright by Powersim Inc., 2009-2011
//
// Date: December 08, 2014 14:10:43
*******************************************************************************/
#include     <math.h>
#include     "PS_bios.h"
typedef float DefaultType;
#define      GetCurTime() PS_GetSysTimer()

interrupt void Task();
void Task_1();

DefaultType  fGblVinA = 0.0;
DefaultType  fGblV_IB = 0.0;
DefaultType  fGblV_IC = 0.0;
DefaultType  fGblVbus = 0.0;
DefaultType  fGblVbusGND = 0.0;
DefaultType  fGblVC = 0.0;
DefaultType  fGblVB = 0.0;
DefaultType  fGblV_IA = 0.0;
DefaultType  fGblVA = 0.0;
typedef struct {
      unsigned long tmLow;
      unsigned long tmHigh;
}     _CBigTime;

_CBigTime GetBigTime(void)
{
      static _CBigTime tm = {0,0};
      unsigned long curTime = GetCurTime();
      if (curTime < tm.tmLow)
            tm.tmHigh++;
      tm.tmLow = curTime;
      return tm;
}

interrupt void Task()
{
      DefaultType fVSAW10, fSIN13, fVDC19, fSUMP18, fVDC29, fSUMP26, fSIN24,
fSUM12, fSUMP25;
      DefaultType fSIN25, fSUMP27, fSIN23, fSIN16, fSIN17;
      PS_EnableIntr();

      {
            static unsigned long period = (unsigned long)(150000000L / 60);
            static float fPeriod = ((float)60) / 150000000L;
            static _CBigTime tmCarrierStart = {0, 0};
            _CBigTime tm = GetBigTime();
            unsigned long tmp1, tmp2;
            tmp1 = tm.tmLow - tmCarrierStart.tmLow;
```

```
            tmp2 = tm.tmHigh - tmCarrierStart.tmHigh;
            if (tm.tmLow > tmCarrierStart.tmLow)
                  tmp2++;
            if (tmp2 || (!tmp2 && (tmp1 >= period))) {
                  tmp1 = tmCarrierStart.tmLow + period;
                  if ((tmp1 < tmCarrierStart.tmLow) || (tmp1 < period))
                        tmCarrierStart.tmHigh++;
                  tmCarrierStart.tmLow = tmp1;
            }
            tmp1 = tm.tmLow - tmCarrierStart.tmLow;
            fVSAW10 = 360 * tmp1 * fPeriod;
      }
      fSIN13 = sin(fVSAW10 * (3.14159265 / 180.));
#ifdef _DEBUG
      fGblVinA = fSIN13;
#endif
      PS_SetPwm1Rate(fSIN13);
      fVDC19 = 120;
      fSUMP18 = fVSAW10 + fVDC19;
      fVDC29 = 180;
      fSUMP26 = fSUMP18 + fVDC29;
      fSIN24 = sin(fSUMP26 * (3.14159265 / 180.));
      PS_SetPwm2Rate(fSIN24);
      fSUM12 = fVSAW10 - fVDC19;
      fSUMP25 = fSUM12 + fVDC29;
      fSIN25 = sin(fSUMP25 * (3.14159265 / 180.));
      PS_SetPwm3Rate(fSIN25);
      fSUMP27 = fVSAW10 + fVDC29;
      fSIN23 = sin(fSUMP27 * (3.14159265 / 180.));
      PS_SetPwm4Rate(fSIN23);
      fSIN16 = sin(fSUMP18 * (3.14159265 / 180.));
      PS_SetPwm5Rate(fSIN16);
      fSIN17 = sin(fSUM12 * (3.14159265 / 180.));
      PS_SetPwm6Rate(fSIN17);
      PS_ExitPwm1General();
}

void Task_1()
{
      DefaultType fTI_DIN1, fNOT16, fTI_ADC1, fVDC20, fSUM15, fIpk_max, fCOMP8,
fTI_ADC1_1;
      DefaultType fSUM13, fCOMP9, fTI_ADC1_2, fSUM14, fCOMP10, fOR31, fTI_ADC1_3,
fABS1, fVbus_max;
      DefaultType fCOMP11, fTI_ADC1_4, fABS2, fCOMP12, fOR1, fTI_ADC1_8, fVDC10,
fSUM7, fVpk_max;
      DefaultType fCOMP13, fTI_ADC1_9, fSUM6, fCOMP14, fTI_ADC1_10, fSUM9,
fCOMP15, fOR32;
      DefaultType fOR33, fNOT18, fTI_TZSTATE1, fNOT9, fTI_DIN1_1, fNOT17, fOR4,
fOR2;

      fTI_DIN1 = (PS_GetDigitInA() & ((Uint32)1 << 24)) ? 1 : 0;
      fTI_ADC1 = PS_GetDcAdc(0);
      fTI_ADC1_1 = PS_GetDcAdc(1);
      fTI_ADC1_2 = PS_GetDcAdc(2);
      fTI_ADC1_3 = PS_GetDcAdc(3);
      fTI_ADC1_4 = PS_GetDcAdc(4);
      fTI_ADC1_8 = PS_GetDcAdc(8);
```

```
	fTI_ADC1_9 = PS_GetDcAdc(9);
	fTI_ADC1_10 = PS_GetDcAdc(10);
	fTI_DIN1_1 = (PS_GetDigitInA() & ((Uint32)1 << 25)) ? 1 : 0;
	fNOT16 = !fTI_DIN1;
	if (fNOT16 > 0)
	{
		PS_StartPwm(1);
	}
	if (fNOT16 > 0)
	{
		PS_StartPwm(2);
	}
	fVDC20 = 1.5;
	fSUM15 = fTI_ADC1 - fVDC20;
	fIpk_max = (150.0/103);
	fCOMP8 = (fSUM15 > fIpk_max) ? 1 : 0;
	fSUM13 = fTI_ADC1_1 - fVDC20;
	fCOMP9 = (fSUM13 > fIpk_max) ? 1 : 0;
	fSUM14 = fTI_ADC1_2 - fVDC20;
	fCOMP10 = (fSUM14 > fIpk_max) ? 1 : 0;
	fOR31 = (fCOMP8 != 0) || (fCOMP9 != 0) || (fCOMP10 != 0);
	fABS1 = fabs(fTI_ADC1_3);
	fVbus_max = (600.0/1365);
	fCOMP11 = (fABS1 > fVbus_max) ? 1 : 0;
	fABS2 = fabs(fTI_ADC1_4);
	fCOMP12 = (fABS2 > fVbus_max) ? 1 : 0;
	fOR1 = (fCOMP11 != 0) || (fCOMP12 != 0);
	fVDC10 = 1.5;
	fSUM7 = fTI_ADC1_8 - fVDC10;
	fVpk_max = (600.0/1365);
	fCOMP13 = (fSUM7 > fVpk_max) ? 1 : 0;
	fSUM6 = fTI_ADC1_9 - fVDC10;
	fCOMP14 = (fSUM6 > fVpk_max) ? 1 : 0;
	fSUM9 = fTI_ADC1_10 - fVDC10;
	fCOMP15 = (fSUM9 > fVpk_max) ? 1 : 0;
	fOR32 = (fCOMP13 != 0) || (fCOMP14 != 0) || (fCOMP15 != 0);
	fOR33 = (fOR31 != 0) || (fOR1 != 0) || (fOR32 != 0);
	fNOT18 = !fOR33;
	fTI_TZSTATE1 = PS_IsPwm1OneShotTZ();
	fNOT9 = !fTI_TZSTATE1;
	fNOT17 = !fTI_DIN1_1;
	fOR4 = (fNOT9 != 0) || (fNOT17 != 0);
#ifdef _DEBUG
	fGblV_IB = fSUM13;
#endif
#ifdef _DEBUG
	fGblV_IC = fSUM14;
#endif
#ifdef _DEBUG
	fGblVbus = fTI_ADC1_3;
#endif
#ifdef _DEBUG
	fGblVbusGND = fTI_ADC1_4;
#endif
#ifdef _DEBUG
	fGblVC = fSUM7;
#endif
```

```
#ifdef _DEBUG
	fGblVB = fSUM6;
#endif
#ifdef _DEBUG
	fGblV_IA = fSUM15;
#endif
	if (fNOT16 > 0)
	{
		PS_StartPwm(3);
	}
	if (fNOT16 > 0)
	{
		PS_StartPwm(4);
	}
	if (fNOT16 > 0)
	{
		PS_StartPwm(5);
	}
	if (fNOT16 > 0)
	{
		PS_StartPwm(6);
	}
	fOR2 = (fOR33 != 0) || (fNOT17 != 0);
	if (fOR2 != 0)
	{
		PS_StopPwm(6);
	}
	if (fOR2 != 0)
	{
		PS_StopPwm(2);
	}
	if (fOR2 != 0)
	{
		PS_StopPwm(1);
	}
	if (fOR2 != 0)
	{
		PS_StopPwm(3);
	}
	if (fOR2 != 0)
	{
		PS_StopPwm(4);
	}
	if (fOR2 != 0)
	{
		PS_StopPwm(5);
	}
#ifdef _DEBUG
	fGblVA = fSUM9;
#endif
	(fNOT18 == 0) ? PS_ClearDigitOutBitA((Uint32)1 << 31) :
PS_SetDigitOutBitA((Uint32)1 << 31);
	(fOR4 == 0) ? PS_ClearDigitOutBitB((Uint32)1 << (34 - 32)) :
PS_SetDigitOutBitB((Uint32)1 << (34 - 32));
}
void Initialize(void)
{
```

```
    PS_SysInit(30, 10);
    PS_StartStopPwmClock(0);
    PS_InitTimer(0, 0xffffffff);
    PS_InitPwm(1, 1, 40000*1, (0.5e-6)*1e6, PWM_TWO_OUT, 10665);    // pwnNo,
waveType, frequency, deadtime, outtype
    PS_SetPwmPeakOffset(1, 1.3, 0, 1.0/1.3);
    PS_SetPwmIntrType(1, ePwmNoAdc, 1, 0);
    PS_SetPwmVector(1, ePwmNoAdc, Task);
    PS_SetPwmTripZone(1, 1, eTzEnOneShot);
    PS_SetPwmTzAct(1, eTZHighImpedance);
    PS_SetPwm1Rate(0);
    PS_StopPwm(1);

    PS_InitPwm(2, 1, 40000*1, (0.5e-6)*1e6, PWM_TWO_OUT, 10665);    // pwnNo,
waveType, frequency, deadtime, outtype
    PS_SetPwmPeakOffset(2, 1.3, 0, 1.0/1.3);
    PS_SetPwmIntrType(2, ePwmNoAdc, 1, 0);
    PS_SetPwmTripZone(2, 1, eTzEnOneShot);
    PS_SetPwmTzAct(2, eTZHighImpedance);
    PS_SetPwm2Rate(0);
    PS_StopPwm(2);

    PS_InitPwm(3, 1, 40000*1, (0.5e-6)*1e6, PWM_TWO_OUT, 10665);    // pwnNo,
waveType, frequency, deadtime, outtype
    PS_SetPwmPeakOffset(3, 1.3, 0, 1.0/1.3);
    PS_SetPwmIntrType(3, ePwmNoAdc, 1, 0);
    PS_SetPwmTripZone(3, 1, eTzEnOneShot);
    PS_SetPwmTzAct(3, eTZHighImpedance);
    PS_SetPwm3Rate(0);
    PS_StopPwm(3);

    PS_InitPwm(4, 1, 40000*1, (0.5e-6)*1e6, PWM_TWO_OUT, 10665);    // pwnNo,
waveType, frequency, deadtime, outtype
    PS_SetPwmPeakOffset(4, 1.3, 0, 1.0/1.3);
    PS_SetPwmIntrType(4, ePwmNoAdc, 1, 0);
    PS_SetPwmTripZone(4, 1, eTzEnOneShot);
    PS_SetPwmTzAct(4, eTZHighImpedance);
    PS_SetPwm4Rate(0);
    PS_StopPwm(4);

    PS_InitPwm(5, 1, 40000*1, (0.5e-6)*1e6, PWM_TWO_OUT, 10665);    // pwnNo,
waveType, frequency, deadtime, outtype
    PS_SetPwmPeakOffset(5, 1.3, 0, 1.0/1.3);
    PS_SetPwmIntrType(5, ePwmNoAdc, 1, 0);
    PS_SetPwmTripZone(5, 1, eTzEnOneShot);
    PS_SetPwmTzAct(5, eTZHighImpedance);
    PS_SetPwm5Rate(0);
    PS_StopPwm(5);

    PS_InitPwm(6, 1, 40000*1, (0.5e-6)*1e6, PWM_TWO_OUT, 10665);    // pwnNo,
waveType, frequency, deadtime, outtype
    PS_SetPwmPeakOffset(6, 1.3, 0, 1.0/1.3);
    PS_SetPwmIntrType(6, ePwmNoAdc, 1, 0);
    PS_SetPwmTripZone(6, 1, eTzEnOneShot);
    PS_SetPwmTzAct(6, eTZHighImpedance);
    PS_SetPwm6Rate(0);
    PS_StopPwm(6);
```

```
        PS_ResetAdcConvSeq();
        PS_SetAdcConvSeq(eAdcCascade, 0, 1.0);
        PS_SetAdcConvSeq(eAdcCascade, 1, 1.0);
        PS_SetAdcConvSeq(eAdcCascade, 2, 1.0);
        PS_SetAdcConvSeq(eAdcCascade, 3, 1.0);
        PS_SetAdcConvSeq(eAdcCascade, 4, 1.0);
        PS_SetAdcConvSeq(eAdcCascade, 8, 1.0);
        PS_SetAdcConvSeq(eAdcCascade, 9, 1.0);
        PS_SetAdcConvSeq(eAdcCascade, 10, 1.0);
        PS_AdcInit(0, !0);

        PS_InitDigitIn(24, 100);
        PS_InitDigitIn(25, 100);

        PS_InitDigitOut(31);
        PS_InitDigitOut(34);

        PS_StartStopPwmClock(1);
}

void main()
{
        Initialize();
        PS_EnableIntr();    // Enable Global interrupt INTM
        PS_EnableDbgm();
        for (;;) {
                Task_1();
        }
}
```

6.3. CODE IMPLEMENTATION AND TESTING

Using the Code Composer Studio software, it was possible to implement the code generated in the DSC TI F28335. As a first step, the waveforms on the switches of the same phase of the converter were checked, only with the power supply to the drivers, i.e. with no input voltage on the DC bus. Figure 114 shows the waveforms between the Gate and Source switches of phase A of the NPCm converter.

Figure 114 - Waveforms between Gate and Source of phase A switches.

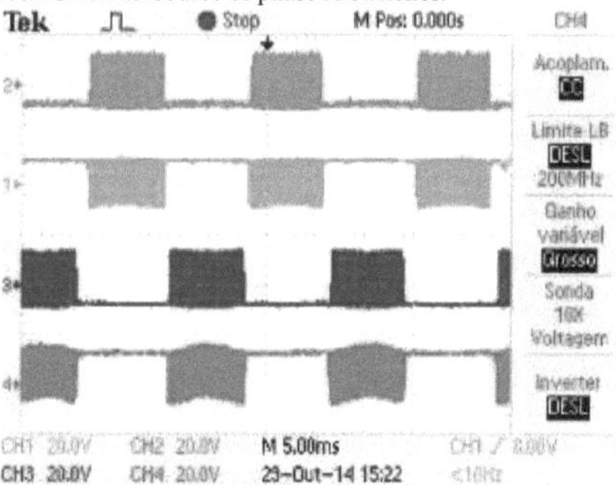

Source: Author's own production.

Channel 1 of the oscilloscope, seen in the previous figure, shows the PWM signal from switch S02, channel 2 from Sp, channel 3 from Sn and channel 4 from switch S01. It can be seen that, as in the simulation, the signals from switches Sp and S02, as well as Sn and S01, are complementary. It can also be seen that the signals from switches Sn and S01 are 180° off in relation to switches Sp and S02. By scaling down the oscilloscope, you can better visualise that the PWM signals from the Sp and S02 switches are complementary, as shown in Figure

115.

Figure 115 - Waveforms on the Sp and S02 switches.

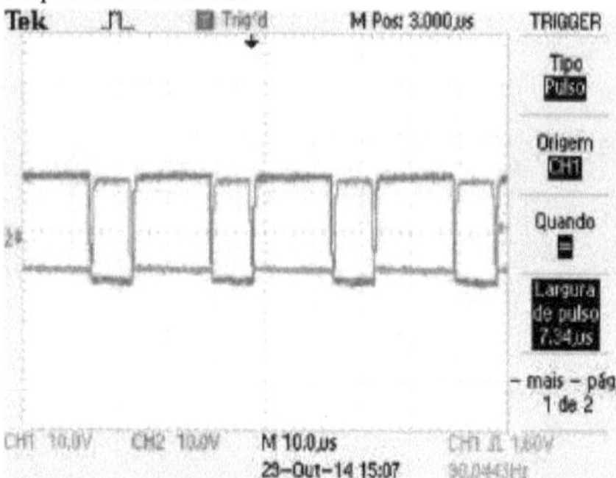

Source: Author's own production.

If you reduce the scale even further, you can see from the oscilloscope's dotted grid that the dead time set in the PSIM simulation of 0).5LIS is verified, as shown in Figure 116.

Figure 116 - Dead time between switches Sp and S02.

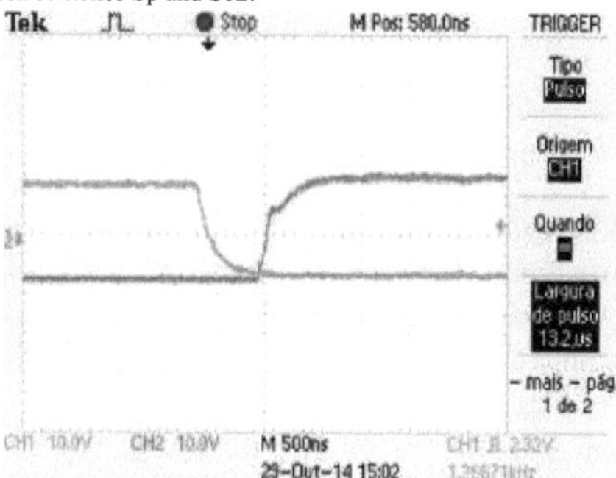

Source: Author's own production.

Another configuration checked was the switching frequency of the PWM channels, using cursors on the oscilloscope. Figure 117 shows the switching frequency, which is 40kHz, just like the one configured in the simulation for generating the code.

Figure 117 - Checking the switching frequency.

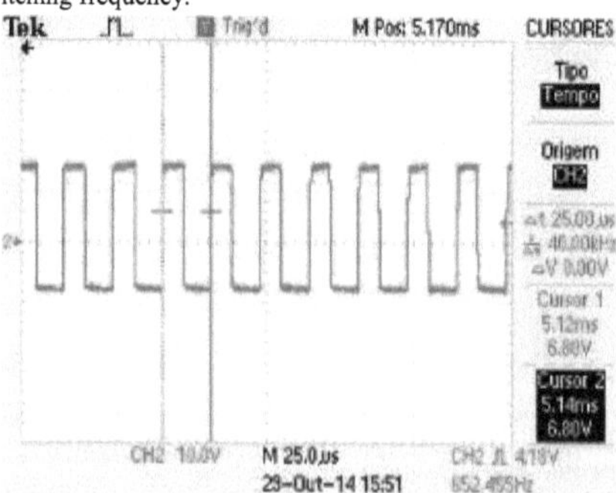

Source: Author's own production.

The frequency of the reference signal was also checked using the oscilloscope cursors. In Figure 118, the four channels of the oscilloscope were used to check the waveforms of the switches, as well as the frequency of the reference signal.

Figure 118 - Checking the frequency of the reference signal.

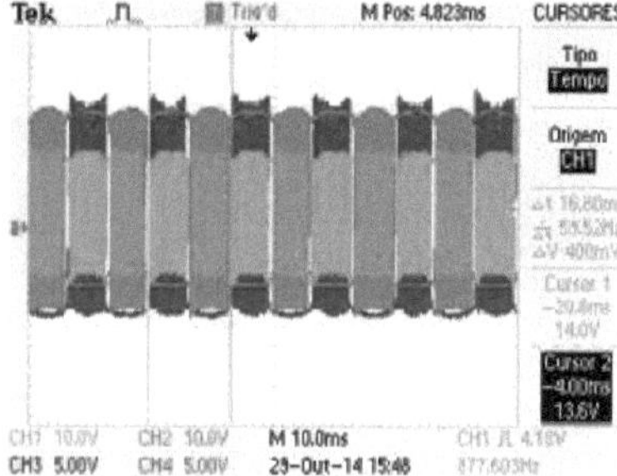

Source: Author's own production.

With the operation of the switches verified, the voltage on the DC bus was then triggered, using a rectifier powered by a three-phase Varivolt. Figure 119 shows the voltage waveform between the Drain and Source of the SpA switch, showing two voltage levels.

Figure 119 - Voltage waveform between the Drain and Source of the SpA switch

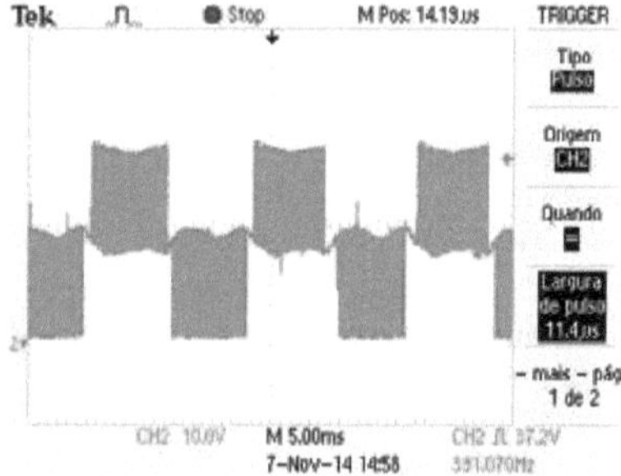

Source: Author's own production.

As a comparison, the waveform resulting from the simulation can be seen in Figure 120. The simulation also shows two voltage levels, just like the ripple found in the waveform obtained experimentally.

Figure 120 - Voltage between Drain and Source obtained in the simulation.

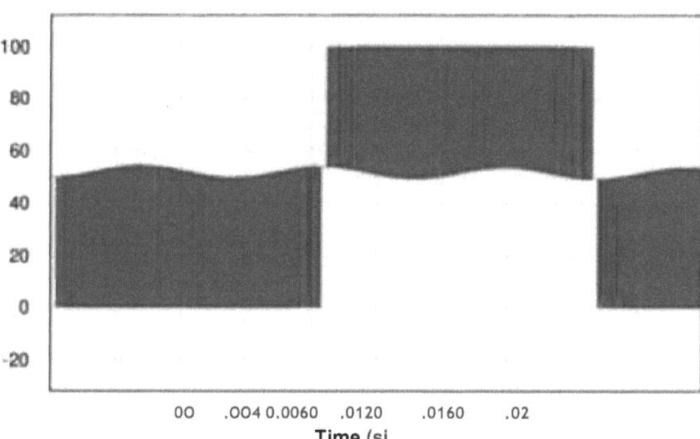

Source: Author's own production.

In the same way that the voltage waveform was obtained for switch SpA, it was obtained for S02A, i.e. between Drain and Source. This waveform is shown in Figure 121.

Figure 121 - Voltage waveform between Drain and Source of switch S02A.

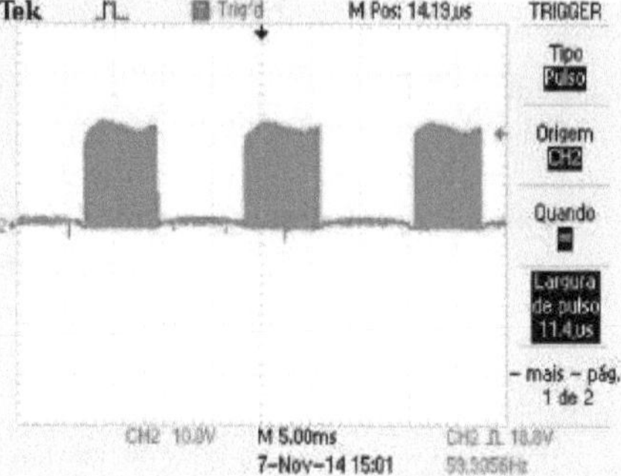

Source: Author's own production.

It can be seen that this switch has a complementary waveform for only half a period of the reference signal, i.e. when the reference signal is greater than zero. Similar waveforms were observed for the SnA and S01A switches, offset by 180° from the waveforms shown above.

Figure 122 shows the waveforms between the Gate and Source of the SpA and S02A switches. It can be seen that there is no dead time between the complementary signals, due to the capacitive characteristics of the switches.

Figure 122 - Checking that there is no dead time between the SpA and S02A switch signals.

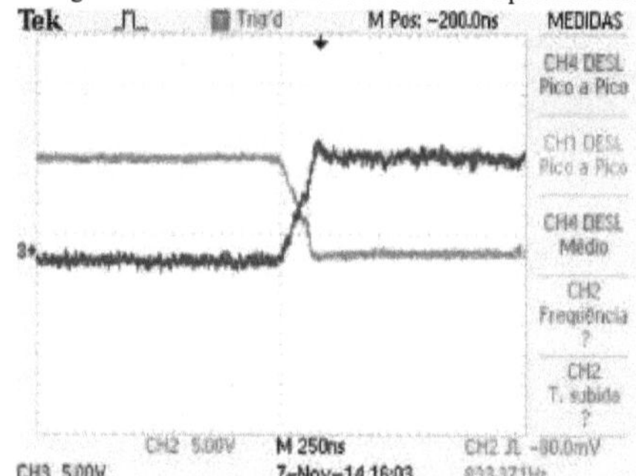

Source: Author's own production.

The DC bus current and voltage waveforms were then obtained. Figure 123 shows the waveforms. It can be seen that the NPCm output current is distorted when it passes through zero, as in the simulation, due to the dead time between the PWM channels on the same arm of the converter. This dead time, in turn, is due to the small offset level of the carrier. It can be seen that the DC bus voltage is 17.6V, while the current has an amplitude of 1.6A.

Figure 123 - DC bus voltage and phase A output current.

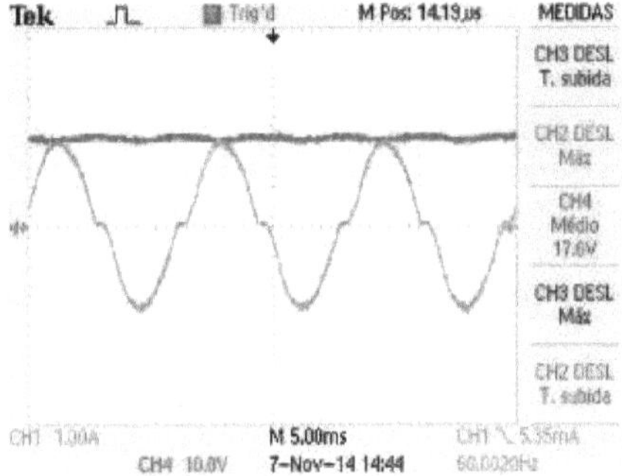

Source: Author's own production.

Figure 124 shows the voltage and current output waveforms for phase A of the converter. It can be seen that the DC bus voltage has been increased, so that the output voltage has an amplitude of approximately 14V, and the current 5A.

Figure 124 - Output voltage and current of phase A of the NPCm.

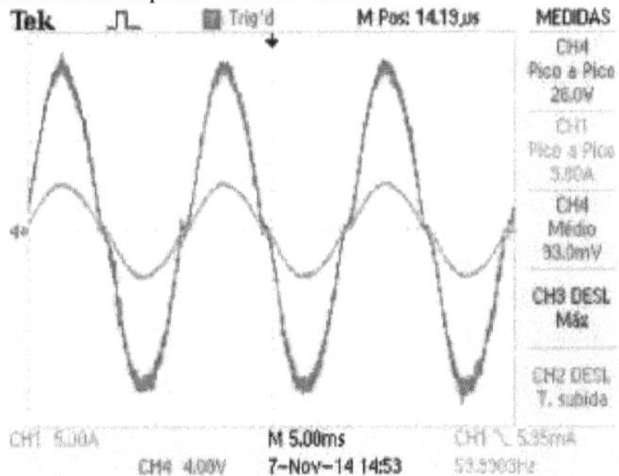

Source: Author's own production.

To show how the three phases work, Figure 125 shows the currents obtained from the converter phases. You can see differences in the waveforms in the phases due to differences in the oscilloscope current pointers used to obtain the data.

Figure 125 - Currents in the three phases of the NPCm.

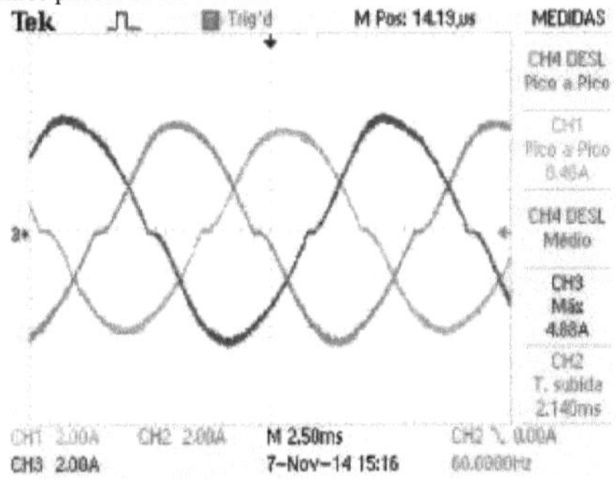

Source: Author's own production.

CHAPTER 7

7. CONCLUSIONS

In this final project, we studied the tools available in the PSIM software® . Using SimCoder elements, it is possible to carry out simulations of power electronics circuits, in which PWM configurations, digital inputs and outputs, A/D converters, among others, are carried out for application in a DSC. PSIM® is capable of generating code in C language for the Texas Instruments TMS320F28335 DSC® , which can be used in Code Composer Studio software without having to modify the code. This work was designed to be a guide for future applications of these tools, presenting the steps for their application. The objectives of this work were listed in Chapter 1.

Chapter 2 studied converters with switches, including the Buck Converter and some inverter topologies. For the Buck Converter, the transfer function of the voltage as a function of the cyclic ratio was obtained. For the inverters, we started by showing the topology with two switches and then increased the number of switches, also increasing the number of phases and consequently the complexity of the converter control.

Chapter 3 introduced the SimCoder blocks, their configuration parameters and how they work. Some of the most commonly used blocks can be highlighted, such as the PWM modulators, the digital inputs and outputs and the A/D converter. Other elements that can be used to generate code are also shown in this chapter.

Chapter 4 showed the process for generating code in C language in PSIM® , and then briefly explained the structure of the generated code. It was noted that CCS automatically updates the generated code after changes are made to the simulation, only if the code file is in the same workspace as the one used in CCS. The process for importing the code generated in PSIM® into CCS was also explained in this chapter. The files generated by PSIM® and the main characteristics of the DSC used, the TMS320F28335, from Texas Instruments® were listed.

In Chapter 5, simulations were carried out with the converters explained in Chapter 2, using the SimCoder configuration blocks to generate codes in C language. The Buck Converter was simulated in open loop and closed loop, with a PI controller, while the DC-AC converters were simulated only in open loop, with sinusoidal references.

Finally, Chapter 6 tests the TMS320F28335 DSC. Simulations were carried out in PSIM® with the most commonly used blocks in order to test them individually. Waveforms were obtained using oscilloscopes and various important characteristics of the PWM modulators, the A/D converter and the digital inputs and outputs were checked. Also in Chapter 6, the NPCm, a three-phase reversible DC-AC converter, was studied. Using the code generation blocks, the converter was simulated. At this stage, some difficulties were encountered with the control, since a converter that was already ready was used and its design specifications had to be followed. The main difficulty was with the PWM blocks, which had to follow the order of the specified channels. However, it was possible to carry out the simulation by modifying the In-Phase Disposition modulation, using only one triangular carrier and using two sinusoidal references for each phase, offset by 180°. The code was then generated in C language and implemented in the DSC found in the converter. Using an oscilloscope, various waveforms were obtained of the command signals on the switches, the current and the output voltage of the NPCm, verifying the functionality of the code generated.

The SimCoder tool proved efficient for configuring the DSC. However, it is still a tool under development and needs to be studied further. A suggestion for future work is to research and test the SCI (Serial Communication Interface) and SPI (Serial Peripheral Interface) blocks, which were not tested because they are only found in version 9.1 of the PSIM software® , and the previous version was used in this work. Another point to be suggested is the practical testing of the control done for the Buck Converter and for other converters.

REFERENCES

AHMED, A. Power Electronics. São Paulo: Prentice Hall, 2000.

BARBI, I. Power Electronics. 6ª . ed. Florianópolis: [s.n.], 2006.

HEERDT, J. A. Three-phase Active Electronic Load. Florianópolis: [s.n.], 2013.

MARTINS, D.; BARBI, I. Basic Non-Isolated DC-DC Converters. 2nd ed. Florianópolis: [s.n.], 2006.

POWERSIM. SimCoder User's Guide. [S.l.]: [s.n.], 2010.

POWERSIM. PSIM User's Guide. [S.l.]: [s.n.], 2013.

TEXAS INSTRUMENTS. Code Composer Studio v5 User's Guide. [S.l.]: [s.n.], 2011.

TEXAS INSTRUMENTS. TMS320F28335 Data Manual. [S.l.]: [s.n.], 2011.

Printed by Books on Demand GmbH, Norderstedt / Germany